Molecular Biology of Plants
A Laboratory Course Manual

Instructors
Russell Malmberg, *Cold Spring Harbor Laboratory*
Joachim Messing, *University of Minnesota*
Ian Sussex, *Yale University*

Assistants
Margaret Boylan, *Yale University*
Michael Zarowitz, *University of Minnesota*
Jean McIndoo, *Cold Spring Harbor Laboratory*

Cold Spring Harbor Laboratory Summer Course
June 8–June 28, 1984

Cold Spring Harbor Laboratory
Cold Spring Harbor, New York 11724

MOLECULAR BIOLOGY OF PLANTS

Other manuals available from Cold Spring Harbor Laboratory

Experiments with Gene Fusions
Molecular Cloning
Advanced Bacterial Genetics
Hybridoma Techniques
Experiments in Molecular Genetics
Experiments with Normal and Transformed Cells

All Cold Spring Harbor Laboratory publications may be ordered directly from Cold Spring Harbor Laboratory, Box 100, Cold Spring Harbor, New York 11724. Phone: 1-800-843-4388 in continental U.S.; (516) 367-8425 in New York State.

TABLE OF CONTENTS

OVERVIEW

INTRODUCTION

This course is intended for people with a molecular biology background who wish to become familiar with plants as experimental organisms. Our goal is to familiarize the student with some of the current research and techniques in the field of plant molecular biology, and also to introduce a modest amount of classical botany and plant physiology. We hope to show some of what makes plant life unique and different from other systems.

Among the many interesting experimental systems in the plant kingdom, we have chosen to focus the laboratory work on Nicotiana tabacum (tobacco) and Zea mays (maize). This choice is made because these materials are used in the instructors' own laboratories, because they represent both monocots and dicots, and because they allow us to cover both cell culture and whole plant genetics. Several other species are used for occasional experiments. You should feel free to take with you any of the strains or cultures used during the course.

Plant experiments often take a long time to finish. The consequences of this on the laboratory schedule are: (i) experiments are grouped with respect to timing, rather than by relatedness of topic; (ii) many days there will be a main experiment to perform plus several continuing experiments to monitor; (iii) some experiments may not get finished during the time of the course, although most of the critical steps will have been performed. On the next two pages we give a lab flow chart, and an outline grouping related experiments, in order to lessen the confusion. Protocols are given in this manual in the chronological order in which they are to be initiated.

Each day will begin at 9:00 AM with a 3 hour lecture/seminar held in the second floor library of McClintock Building (brick and stucco building directly down the hill behind the Blackford Cafeteria). After lunch, about 1:30 PM, the laboratory sessions will begin in the top floor laboratory of Delbruck Building (the gray and green shingle building on Bungtown road just beyond the Gazebo). In the evenings we will either continue with the lab, have discussions related to the experiments, or have additional talks.

This is the fourth year of the Plant Course at Cold Spring Harbor Laboratory. The first two years (1981, 1982) the instructors were John Bedbrook and Fred Ausubel; the third year (1983) they were John Bedbrook, Russell Malmberg, and Ian Sussex. The first 3 years of the course were supported by a grant from the National Science Foundation, and this grant has been renewed for 3 more years (NSF-PCM-8316292). A major factor in the renewal of the grant (and the course) was the comments of the students of the first 3 years. These comments also influenced the current choice of experiments and seminar speakers. We strongly encourage you to submit course criticisms to us when it is over.

We thank Regina Schwarz (Delbruck Building), Mike Ockler (Art) and Nancy Ford and Judy Cuddihy (Publications) for helping us put this manual together; we thank Gerry LoFranco (Delbruck) for her help before and during the course.

The experimental protocols given in this book have been contributed by a variety of workers in the plant sciences. We have tried to attribute the origin of each procedure where possible; and we would like to apologize to any contributor or originator of a technique whom we have inadvertantly not identified. We are especially grateful to Burle Gengenbach, Ron Phillips, Andy Wang, Steve Dellaporta, Paul Chomet, and Don Miles for their work in designing and leading specific laboratory sessions during the 1984 course, and the Maureen Hanson for writing the introduction to plant cell culture appendix.

The plant course would not have been possible without the work contributed by the laboratory assistants who helped plan and prepare the laboratory sessions: Margaret Boylan, Susan Brown, Barbara Dunn, Jonathon Jones, Jean McIndoo, Rod Reidel, Vinni Schoene, Jane Smith, Stephen Smith, Kit Steinbeck, John Waldron, and Mike Zarowitz.

CHRONOLOGICAL LABORATORY FLOW CHART

Primary Experiment	Secondary and Continued Experiments	
F 8: Anatomy & Development		
S 9: Anatomy & Development		
S10: Tobacco Cell Culture		
M11: Maize Embryogenesis	LT Selection	
T12: Grafting Tomato	Crown Galls	
W13: Tobacco leaf protoplasts isolation and fusion		
R14: Maize genetics	Maize minipreps	Seeds to Culture
F15: Maize genetics	Field day / pollination	
S17: Tobacco culture RNA		
M18: Co-cultivation	Tobacco RNA	
T19: Maize cytogenetics	In situ hybridization	
W20: Maize cytogenetics		Co-cultivation
R21: Tobacco leaf protoplast	Liposome preparation and fusion	
F22: Maize mt DNA	CAT assay	Suspen. Transfer
S23: Maize genome DNA	Maize mt DNA	Co-cultivation
S24: Nitrate reductase assay	SDS Protein Extract	Maize genome DNA
M25: Maize photosynthesis		
T26: Maize photosynthesis	Harvest Bean Embryos	
W27: Bean embryo rocket	Evaluate cultures	

EXPERIMENTS GROUPED BY TOPICS

PROTOCOLS

PLANT MORPHOLOGY AND ANATOMY LABORATORY

Introduction

The study of plant (and animal) structure has been influenced by biologists having a comparative or evolutionary outlook. Thus there has developed a "type" system for thinking about the relationship between the structures of different kinds of organisms. There is considered to be a normal type of organism, organ, tissue, etc., from which various specializations can occur. These specializations can be considered to be:

1. <u>Modifications</u> of the normal type with no evident special function.

2. <u>Adaptations</u> to particular evolutionary/ecological situations. In plants many of these adaptation are to special environments, and many are concerned with the water economy of the plant. These may be surface modifications or alterations in surface/volume relationships or physiological changes.

3. <u>Aberrations</u> where the normal development and structure are altered as a result of interactions with some other organisms.

Each of the 3 principal organs of the plant, stem, leaf and root, may be specialized from the normal type in any of these three ways. This lab is planned to explain the so-called normal type of structure and some of its specializations.

A. <u>Study of a "normal type" of plant and some modifications.</u>

1. <u>Normal type.</u> Take one of the tomato plants that are available and use it as an example of a normal type of plant. Be careful to remove it with minimum disturbance to the root system. The following is a list of information that is useful to have about a plant and forms the basis for studies about its development. This information will also be used for examining other plants and determining in what way their structure and development differs from the normal type.

<u>Stem</u> How are leaves arranged at the nodes, 1 or more per node?

What is the phyllotaxy, opposite or alternate?

Are axillary buds visible? Is their growth inhibited or not?

Are internodes elongated or not (rosette plant).

What is the distribution of elongation of internodes? What is the plants "architecture?")

<u>Leaf</u> Is the leaf simple or compound?

Is it differentiated into blade, midrib, and petiole?

7

What is its symmetry, dorsoventral, radial, or bilateral?

Does leaf shape change with position on the plant?

Are the leaf surfaces smooth or hair-covered?

Is the leaf venation parallel or net?

Root   How are lateral roots arranged on a root, two, three, etc. rows?

Are lateral roots determinate or not?

Does the primary root persist or not?

Do any, or all, roots have cambial activity?

2.   Modifications.   Examine some of the plants in this collection and identify in what way the stem, leaf, or root is modified in its development from what you would consider to be normal.   The modifications here do not seem to be adaptations or aberrations and simply indicate the range of structural diversity that might be expected to be encountered in large groups of organisms.

3.   Adaptations.   Make the same kinds of analysis as you did above for the plants in this group.   Remember that many adaptations are for water economy, others are for storage, others are for light gathering or shedding, others are for predation or protection.

4.   Aberrations.   Determine how the normal development of the plant has been altered by the interaction of the plant's developmental system with the invading organisms in each of the cases in this collection.

B.   Anatomy.

As an example of internal structure and the particular pattern that is made, examine the lignified cells of the xylem in the leaf and stem.

1.   Stem.

Cut razor blade hand-sections transversally and longitudinally at the internode and node of the tomato plant.   Identify the different lignified cell types of the xylem.   Locate the primary vascular tissue (consisting of xylem and phloem); is it a continuous band or in separate bundles?   Is there evidence of cambial activity producing secondary tissue?   Observe the vascular connections between the stem and leaf in the nodal sections. Compare your sections with the diagrams of stems and identify as many structures as possible.

Procedure for Making Hand-Sections

8

a. Cut a section of stem from the tomato plant in an internodal area. The part to be sectioned must be held firmly between your thumb and index finger. Make a preliminary cut at a right angle to the long axis of the part approximately 1/2 to 1/4 inches from the end of your thumb and finger. This and subsequent sections should be made by cutting toward you with a slicing motion (as opposed to a chopping motion) of the razor blade, making use of as much blade surface as possible.

b. Make a series of sections (slices), trying to make each as thin as possible. Place each section in a petri dish of 50% ethanol which is used to both kill and fix the cytoplasm of each cell. Although the ideal hand-section should be of the entire part (an entire cross section), keep even those sections which are not entire--frequently they will offer the thinnest pieces of tissue. Make approximately 10 sections. Leave the sections in the ethanol for 4-5 min.

C. Meristems

The shoot apical meristem of Lupinus albus (White lupin, lupini bean) is one of the easiest to dissect and observe. Get some of the lupin seedlings and, paying attention to the phyllotaxy, remove leaves with a fine pair of forceps to reveal the shoot apical meristem. Use a dissecting microscope, and increase the magnification as needed. The shoot apical meristem is a bright green hemisphere between the youngest leaf primordia. What are its approximate dimensions? What is the phyllotaxy? When do leaflets become evident on the young leaf primordia? What is the direction of leaflet initiation? When do hairs first appear on leaf primordia? Where do they first appear?

As time permits make comparable studies of other materials that will be available. Especially dissect other meristems.

9

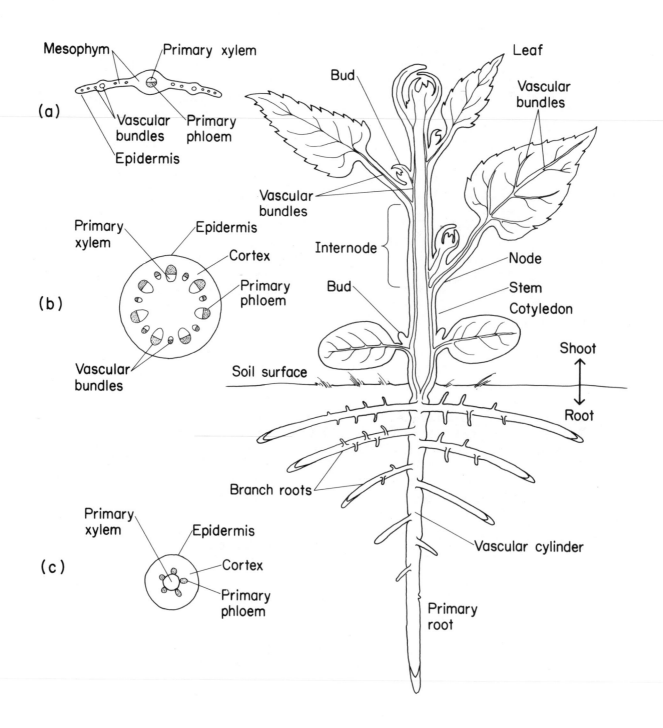

(a)

Mesophym   Primary xylem

Vascular   Primary
bundles   phloem

Epidermis

(b)

Primary
xylem   Epidermis

               Cortex

               Primary
               phloem

Vascular
bundles

(c)

Primary
xylem   Epidermis

               Cortex

               Primary
               phloem

Bud   Leaf

               Vascular
               bundles

Vascular
bundles

Internode

Bud   Node

               Stem
               Cotyledon

Soil surface   Shoot

               Root

Branch roots

               Vascular cylinder

               Primary
               root

Typical Dicotyledonous Plant

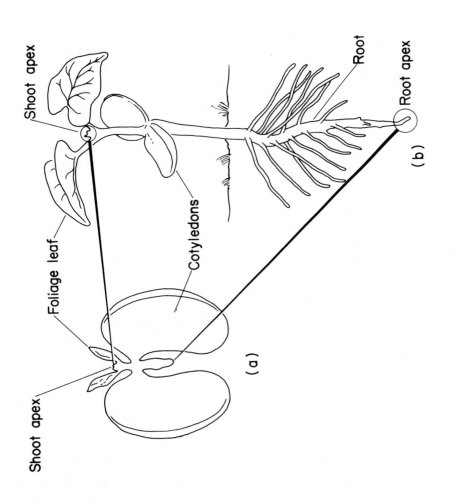

Shoot apex

Foliage leaf

Shoot apex

Cotyledons

(a)

Root

Root apex

(b)

Apical Merisems in the Seed and Seedling

EMBRYOS, MERISTEMS, AND EPIDERMAL HAIRS

Embryos

Maize.   Get an ear of maize, remove the husks, and look at the attachment of the silk (style) to the kernal. Break the ear in half and remove individual kernels (fruit) from the ear. Identify the upper surface by the presence of the embryo. Use a sharp scalpel to make a V-shaped cut in the fruit wall around the embryo. Lift up the flap of tissue to expose the embryo. Lift out the embryo onto a microscope slide and observe it under a stereomicroscope. Look on the upper surface of the embryo for the root-shoot axis, and identify the large boat-shaped scutellum (equivalent to the cotyledon of the dicot embryo).

Bean   The pod is the bean fruit. Identify the stigma and style at the distal end of the pod, and the sepals at the near end. Open up the pod by splitting it longitudinally to reveal the developing seeds. Pull one seed off and identify the micropyle (the hole that the pollen tube grew through to fertilize the egg). It is located near the attachment point of the seed to the pod. Split open the seed coat and remove the embryo. Identify the two cotyledons, and the embryo axis with the shoot apex, and primary leaves, and the root pole at the opposite end of the axis.

Crucifer.   Split the pod (=fruit) lengthwise to reveal the developing seeds. Remove a seed and open it using sharp forceps. Remove the developing embryo. By removing embryos from pods at different positions along the stem of the plant you can see a developmental succession of the embryos.

Meristems

Lupin.   Get some of the germinated seedlings of lupin. Hold them vertical and remove successively younger leaves that are arranged in a spiral phyllotactic sequence. When you cannot remove any smaller leaves put the shoot tip under the stereoscope and, keeping it vertical, remove younger leaves until you get to the shoot apical meristem. Note the shape changes as the leaves develop. The youngest are simple in outline, then leaflet primordia begin to develop and hairs emerge from the leaf surfaces.

Cauliflower.   Break a piece of the white head off the cauliflower and observe it under the stereomicroscope. The whole surface is a mass of vegetative meristems surrounded by leaf primordia. At a later growth stage these meristems will be converted into flower meristems.

Rhododendron.   Remove developing bud scales from the apical bud, at first without using the microscope, then as you get to smaller bud scale under the microscope, holding the shoot vertically. The meristem is surrounded by small developing leaf primordia.

Hairs

Tomato.   In tomato several mutants are known for hair type. Get a wildtype tomato plant and slice off thin epidermal sections of the stem and leaf to identify the number of different hair types that are present. Then look at each of the four mutants to determine how they differ from the wild type.

CELL CULTURE TRANSFERS

Experiments Performed In The Hood

A. Liquid Suspension Cultures
B. Callus Cultures
C. Shoot Cultures
D. Seed Sterilzation
E. Culture Initiation
F. Regeneration From Leaves

Experiments Performed At Benches

G. Sterile Plants To Soil (demonstration)
H. Anther Culture
I. Staining Cultures for Viability
J. Staining Cultures for Chromosomes

Introduction

These protocols are designed to introduce you to many of the standard techniques of plant cell, tissue, and organ culture. This topic includes a wide variety of practices (little theory), whose common denominator is that the material is sterile. Plant cultures generally are grown on simple, defined media (see the Supplement). Plant cultures grow slowly, so that almost any contaminant will out-grow them; hence all work is done in laminar flow hoods with filtered air. Before beginning work, you should:
(1) Turn on the hood about an hour ahead of time.
(2) Clean hood with an aqueous disenfectant cleaning solution (dilute benzalkonium chloride is useful).
(3) Surface sterilize by rubbing the surface with ethanol.

Some useful points to practice are:
(1) Be aware of the direction of airflow, and try not to position objects between the onrushing air, and the sterile material.
(2) Never carry anything open around the room. Always keep petri plates either in their original bag or wrapped with parafilm.
(3) Use sterile disposable plastic pipets for manipulations, but do not mouth pipet. Always open the pipet bag only in the hood, then gently shake them out until one emerges that can be removed without touching others. Reseal the bag when done.
(4) Be wary of drips of media. These can become paths for contaminants to enter the sterile cultures. Wipe drips from flasks, and flame the rims.
(5) Periodically autoclave metal forceps and scalpels, perhaps once per week.

The laboratory we are using has 4 laminar flow hoods for sterile work, 3 up on the top floor, and 1 down below near the kitchen area. Each of these has room for 2 people. Each hood must therefore be used alternately by 2 lab groups. The second group of experiments should be performed when a hood is not available.

Liquid Suspension Cultures

This experiment is intended to be coordinated with the nitrate reductase assay experiment. Each lab group will choose one of 3 media to maintain its liquid suspension cultures, which will then be harvested to test the effects of media on nitrate reductase activity. The media are all based on MS1 (see supplement):

normal MS1
MS1 with no ammonia, only nitrate
MS1 with no nitrate, but with glutamine

The suspension cultures will be a line of <u>Nicotiana tabacum</u> Xanthi named NTX-282. You will need to subculture them every 6 days, judging the necessary dilution based on their rate of growth. This will allow for 2 days of growth prior to harvesting for the nitrate reductase assay.

Maintenance

To subculture a liquid suspension culture, simply take a flask with medium, and a flask containing cells. Flame the rims. Pour the medium into the cells, swirl, and pour the 1:1 mixture back into the empty flask. Wipe any drips with a Kimwipe, and flame the rims. If you repeat this procedure you end up with a 3:1 split. Generally 1:1 and 3:1 are the usual dilutions, although some cultures can readily withstand much greater dilution.

Suspension cultures require constant shaking. They can sit in a hood for manipulation for about a half an hour without damage. If you have your cultures out longer than that, you should swirl them every 10 min.

Experiments that require growth curves in liquid suspension cultures can be performed in a variety of ways. Some finely divided cultures can be monitored by spectrophotometer or Klett meter. Another alternative is to pour the liquid suspensions into a graduated centrifuge tube, and then let them settle to the bottom for 15 to 30 min. The settled cell volume is then read, and the culture returned to the flask. Settled cell volumes may be harmful to some cultures, so the best practice is to pick the shortest settling time that gives good values, and to only measure every 2 to 3 days.

Initiation

To initiate a suspension culture, pick a plate of rapidly growing, friable callus. Inoculate about a half a plate full of callus into a flask containing medium. Shake the suspension culture as in maintenance. After a week, check the culture. It may be possible to subculture it by a 1:1 split, or it may need more time to break apart and begin growth. Avoid subculturing it too dilute. Every week as you subculture it, the larger clumps will gradually disappear (if it is tobacco). Some researchers prefer to filter out the larger clumps through sterile cheesecloth.

Plating

We will actually perform a plating during the co-cultivation experiment. Pour a suspension culture into a graduated 50 ml centrifuge tube, and let settle 15 to 20 min. Remove the supernatant. Add an equal volume of melted, lukewarm, agar medium. Mix, and then pipet 10 ml onto a standard media petri plate. Quickly swirl the plate to make sure the agar-cell slurry is spread evenly. Let the plate sit for 30 min at room temperature to be sure it is hard, then wrap with parafilm and incubate in the dark.

Callus Cultures

Undifferentiated cultures grown on agar medium are generally named callus cultures because of their resemblance to the callus that grows on a whole plant as a wound response. These cultures can be highly undifferentiated - soft, white, friable - or they can be hard rock-like lumps with varying shades of red and brown. Callus can assume particular morphologies, distinguishable shapes, which can be selectively propagated. Success in regenerating some species into whole plants from callus cultures has depended upon recognizing which morphologies are totipotent, and should be maintained.

This experiment simply consists of transferring some pieces of tobacco callus cultures on to new plates. This operation is usually performed every 3 to 4 weeks to maintain stocks. In the hood, unwrap the petri plates with callus in them, and remove several petri plates with fresh medium.

Before attempting to transfer callus, practice folding and wrapping the new plates with parafilm. It takes a few tries to get the right tension. There are doubtless many techniques for doing this, but we routinely cut pieces into rectangles the size of two squares, and then fold this in half lengthwise, and wrap it around the plate.

Use forceps to transfer small hunks of callus to the new plates. There is a minimum size of inoculum, so try several different sizes. Wrap up the plates and put them in the dark incubator. Some species do better as callus in the light, but tobacco seems to prefer the dark. Monitor the growth of the callus every week.

Experiments that require growth curves can be performed with callus by measuring their fresh weight once a week. This is usually done by setting up a top loading balance inside the sterile hood. Callus pieces are temporarily removed from their agar medium, and placed inside a tared petri plate. The combination is then weighed, and the callus returned to its medium.

Shoot Cultures

Shoot cultures are one of the best ways of preserving materials, or genotypes. Of course seeds are better, and require less effort, but shoot cultures can be maintained of stocks that will not flower - such as lethal albino mutants. Shoot cultures also are a ready supply of sterile material, and do not lose the ability to regenerate with time, as do callus cultures. Shoot cultures are maintained in the light on medium that has no hormones.

The essence of shoot cultures is to maintain pieces of stem that have nodes. New shoots will grow forth from buds that exist in the leaf nodes. On an old shoot culture, you can remove most of the leaves and roots, just so each stem section has at least one leaf node.

Take plates of shoot cultures, dissect out stem sections with nodes, and transplant these to deep dish plates with fresh MSO medium. Wrap with parafilm and place in the lighted incubator. Place a stem section on to MSO in a Magenta GA7 vessel, cover, and place in lighted incubator. Monitor growth every week.

15

Two vessels are available on the market, that are larger than deep petri plates, for sterile cultures. Flow Laboratories sells disposable "Plant Cons" in which the agar medium is poured into presterilized containers; and Magenta Corporation (Chicago, IL) sells "GA7" vessels which are reuseable. The Magenta vessels can be autoclaved with medium in them, which may reduce contamination, but also requires washing and recycling. We find the lids to the "Plant Cons" make nice high humidity chambers (see transfer of sterile plants to soil).

STERILIZATION OF MATERIAL

Notes: In sterilization of plant material, a distinction must be made between growth chamber-grown, greenhouse-grown, and field-grown material. The bleach/ethanol treatment described below is basically designed for growth chamber-grown plants, which are normally relatively clean already. Greenhouse-grown plants and field-grown plants are progressively dirtier and more contaminated at the beginning. The sterilization procedure can be made stronger by increasing the length of time in bleach and ethanol, or increasing the concentration of bleach. Obviously, at some point you begin to kill the plant as well. If you have field-grown material which will not sterilize without killing the plant as well, the only solution may be to grow a generation in the cleaner environment of a growth chamber.

Antibiotic media can be used for 2 or 3 months to suppress contaminants. Carbenecillin (1 gm/L) or gentamycin (50 mg/L), and nystatin (500,000 u/L), are useful supplements.

The protocol is useful for enhancing the germination of seeds in addition to sterilizing them.

Sterile Beakers and Flasks   Sterile H_2O

100% Ethanol   10% Bleach and 0.1% SDS

1. Mix plant material with bleach/SDS. Stir occasionally for 5 min. Pour off bleach/SDS, add sterile water. Stir gently and pour off water. Repeat 2 more times.

2. Add 100% ethanol to material, stir, and pour off immediately. Rinse 3 times with sterile water.

3. Put material on appropriate medium. For seeds and shoots, use MS salts with no hormones. For direct callus culture use medium with hormones. Wrap plates in parafilm and incubate in light or dark.

INITIATION OF CULTURES FROM BEAN EMBRYOS

1. Sterilize some bean seeds using the protocol just listed. Place them on MS0 medium, wrap with parafilm, and put in a dark incubator.

2. Monitor the seeds every day. They should swell, and then a small root tip will emerge from the seed. When you can see this, they are ready for dissection.

3. Put the germinated seed in a sterile petri plate in the hood. Using forceps and scalpel, pry off the outer layer (testa). Try to avoid damaging the area near the emerging root. You should be able to see the embryo nestled in between the cotyledons. Cut and remove the cotyledons from the embryo.

4. The goal is to dissect the embryo into cylindrical slices, and then place these slices onto callus medium (MS1), in the same order they are on the embryo. This will allow you to see which parts of the embryo give rise to the best callus. Cut 2 mm slices across the embryonic axis, beginning with shoot apical end, then line these up on the callus medium plate.

5. Wrap up the plates with parafilm, and place in the dark. Check them every few days.

REGENERATION OF TOBACCO PLANTS FROM CULTURE

Regeneration of plants from callus can occur when cultures are shifted from a balance of auxin/cytokinin, and shifted to medium containing high levels of cytokinin, and incubated in the light. Tobacco does this relatively readily, which is why it is the standard model system. The capacity to regenerate is often lost with time, and under the best of circumstances can easily take 2 to 3 months. The strategy is to check the cultures every 3 to 4 weeks, and then to selectively propagate the greenest most organized regions. Well-formed shoots are then transferred to a rooting medium (either no hormones, or just a little auxin). Rooted plantlets can be transferred to soil, grown in larger containers, or propagated as shoot cultures.

In this laboratory we would like you to take some pieces of sterile leaves, and place them on deep petri plates containing MS4 (a regeneration medium), and then incubate in the light incubator. Towards the end of the course we should see some indications of regeneration. You can also try putting some pieces of callus on regeneration medium, and then take them home with you to follow the course of the experiment.

Sterile Plants To Soil (demonstration)

This will be a demonstration by an instructor or assistant. The only real trick to successful transplanting is to maintain high humidity at first, and then to gradually acclimate the plants to the lower humidity of the growth chamber or greenhouse.

Take a sterile shoot culture with roots, and add a layer of water over the agar. Partially break up the agar with a scalpel, but avoid damaging the roots. Let this sit for 30 min to overnight. Remove the plant and remaining agar by scooping it up from below. The agar can be removed by holding the plant inverted under a gentle stream of water, and gradually brushing the agar away. Next, hold the plant in a pot, and pour soil in around the roots. Water the plant. Arrange some sort of covering which should not touch the leaves themselves, and transfer to the growing area. After several days, remove the cover for a short period of time to begin hardening off the plant. Gradually remove it for increasing periods of time.

We find that the lids to Flow Laboratories' "plant cons" make ideal covers for maintaining high humidity, provided the plant is in a rectangular pot with about a 9.5 cm side at the top.

ANTHER CULTURE

Normally this is a sterile procedure, but we will perform it out in the lab. Anther culture normally takes 4 to 6 weeks for a result; we hope to have some set up ahead of time that will pop open during the course. Anther culture is a relatively easy way of making haploid plants of tobacco, but with some other species they result in dihaploids, plantlets that are completely homzygous diploids.

Anther culture can be performed by carefully dissecting out the anthers individually, and culturing them, or by an easier method which has lower frequencies of success.

Remove floral buds from a tobacco plant. The ideal time is about when the corolla is just growing past the sepals, but you should take several buds of different ages around this stage. Hold the bud at the base with thumb and finger, and then dip in ethanol to sterilize (we assume that the interior of the flower is sterile). With forceps, remove the tip of the bud, making an opening to the interior. Hold the bud over the petri plate with medium, with the opening down. Gently rotate the bud with your fingers, and gradually squeeze out the anthers, without touching them, so that they drop onto the plate. Do not move the anthers from their position. You can remove any other flower parts that may have landed on the plate. Incubate in low light for 4 to 6 weeks.

The anthers should turn a uniform brown, and show no bruises. If it works, they will suddenly pop open, and many tiny plantlets will emerge. Tease these apart and grow as a sterile shoot culture. To get haploid callus, it is better to take younger material near the shoot tip, as polyploidization can occur during plant development. You should start 5 to 10 independent haploid calli, in order to have a good chance of getting a stable suspension culture that is haploid.

A cold pretreatment of the buds can enhance the frequency. Put the floral buds at 4°C for 1 to 3 days prior to culture.

STAINING CULTURES FOR VIABILITY

<u>Needed</u>: 0.1% Bromphenol Blue
Tobacco Cell Culture
Microscope

A variety of dyes exist that are excluded by living plant cells, but that stain dead cells. Bromphenol blue can work like this.

1. Take a small aliquot of cells (2 ml) out of a liquid suspension culture of tobacco. Remember to do this in a sterile manner. Split the cells into 2 tubes. Put one tube at high temperature (65°C to 100°C) for a few minutes to kill the cells. Make up 3 samples:

    100% untreated (normal) cells
    50:50 normal and heat killed
    100% heat killed

2. Mix an equal volume of 0.1% Bromphenol blue with each of the 3 samples. Wait 2 to 5 min, and then pipet a few cells on to a slide. Put on a cover slip, and examine the cells under the microscope under low power.

3. Count the percentage of blue cells for each sample.

Wild type cell cultures will vary in the percentage cells alive throughout the growth cycle. A higher percent will be alive in exponential vs. stationary phase. Thus, when measuring cell death due to some treatment, it is necessary to keep the growth phase in mind.

STAINING CULTURES FOR CHROMOSOME COUNTS

1. Pretreatment:
   a) quickly dividing cultures - 50% saturated orthodichlorobenzene, 1 hr at room temperature or 0.01% colchicine, 4 hr at room temperature.
   b) slowly dividing cultures - 0.05% colchicine, 4 hr at room temperature.

2. Fixation:
   3 parts 95% ethanol: 1 part glacial acetic acid (Farmer's Fixative), or 6 parts Ethanol: 3 parts chloroform: 1 part acetic acid (Carnoy's Fixative). Overnight at 4°C.

3. Storage:
   70% ethanol at 4°C, for up to 5 days.

4. Maceration (optional):
   0.5% cellulase + 0.5% pectinase in 0.1 M Na acetate, pH 4.5
   2 hr at room temperature, then wash with 70% ethanol.

NOTE: In the course, you will be given cells already treated as above.

5. Staining:
   Place cells on slide, let dry. Add a few drops of MCF or PropioCarmine Stains. If you are using PropioCarmine, mix in some iron (a rusty scalpel is useful) and gently warm slide. Put on cover slip and squash.

Carbol Fuchsin A	Carbol Fuchsin B
3 gm basic fuchsin	10 ml of CF-A
100 ml 70% ethanol	90 ml 5% phenol
stores indefinitely 4°C	stores 2 weeks 4°C

CF-Stain	MCF Stain
45 ml of CF-B	10 ml CF Stain
6 ml acetic acid	90 ml 45% acetic acid
6 ml 37° Formaldehyde	1.8 gm sorbitol
stores indefinitely 4°C	age 2 weeks room temp
	stores room temperature

PropioCarmine

0.5% carmine
45% propionic acid

MAIZE EMBRYOGENIC TISSUE CULTURES

I. Culture Initiation

    A. Embryo isolation (see Appendix).

        1. Make desired controlled pollination.

        2. Take ear when embryos are at appropriate size (usually about 1.5 mm long).

        3. Remove husks and surface sterilize in 2.5% sodium hypochlorite + 1-2 drops detergent for 20 min; rinse 3 times with sterile H_2O.

        4. Slice off top of kernels; remove endosperm with embryo usually attached; separate embryo.

        5. Place embryo with axis side down onto agar solidified media.

    B. Initiation media (Armstrong and Green, in prep.)

        1. N6 media (see Supplement).

        2. Add 20 g/l sucrose; 6 mM L-proline; 100 mg/l casamino acids; 1 mg/l 2,4-D; pH to 5.8.

        3. Add 7 g/l agar (0.7% final) and autoclave.

        4. Pour into 25 x 100 mm Petri dish, etc.

    C. Incubation

        1. Temp. 28-30°C.

        2. Light - cool white fluorescent 16:8 h light: dark cycle 2-20 $umol/m^2/s$.

II. Culture Maintenance

    A. Media same as for initiation.

    B. Subculture frequency at 10-14 days.

    C. Conditions to favor maintenance of friable embryogenic cultures include incubation in low light to reduce differentiation.

    D. Asceptically transfer healthy culture inocula of pea to dime size to fresh media.

III. Plant regeneration.

    A. Identify culture sectors containing somatic embryoids at or near the coleoptilar ring stage.

23

B. Transfer to N6 medium lacking 2,4-D but with 6% sucrose.

C. After about 14 d transfer the "mature" embryoids to MS medium with 2% sucrose and no 2,4-D.

D. Embryos will "germinate" and produce leaves, shoots and roots.

E. Transfer plantlets to soil: vermiculite mixture and grow at medium light and high humidity for 14 to 21 d.

F. Transplant to field or greenhouse.

IV. Selection with friable embryogenic cultures.

A. Establishment of the effective treatments for selection.

1. For LT (lysine + threonine) selection, prepare N6 maintenance medium with added equilmolar L-lysine and L-threonine ranging from 0.1 mM to 5.0 mM.

2. Transfer friable embryogenic culture inocula to LT medium and incubate at standard conditions.

3. Observe cultures after 1 and 2 weeks for growth inhibition (for accurate data weigh the culture inocula before and after the growth period; this can be difficult for these cultures).

4. Transfer cultures to similar LT media for second growth period; some of the LT concentrations may be eliminated if they have no effect; there can be more inhibition during the second selection cycle than was apparent during the first.

B. Deciding on selection strategy.

1. Lethal dose or go for broke - need a lot of culture material because all or most of it will die; pick LT concentration shown to be very inhibitory and transfer a lot of cultures onto it.

2. Sublethal enrichment or easy does it - slower approach used when culture material is limiting; pick LT concentration shown to inhibit culture growth by about 50-75%; repeated transfers to LT media at the initial and increasingly higher concentrations are required.

C. Characterization of resistant cultures.

1. Split culture and grow in absence of LT for several cycles and then rechallenge with lethal LT.

2. Regenerate plants, cross onto control plants and/or self pollinate, then retest immature embryos, etc. for LT resistance.

References:

Armstrong, C.L., Green, C.E. (1982) Initiation of friable, embryogenic maize callus: the role of L-proline. Agron. Abst. p. 89.

Green, C.E. (1982) Somatic embryogenesis and plant regeneration from the friable callus of _Zea mays_. In: Plant tissue culture 1982, pp. 107-108, Fujiwara, A., ed., Maruzen Co. Ltd., Tokyo.

Green, C.E. (1983) New developments in plant tissue culture and plant regeneration. In: Basic biology of new developments in biotechnology, pp. 195-209, Hollaender, A., Laskin, A.I., Rogers, P., eds., Plenum Publishing Corp., New York.

Green, C.E., Armstrong, C.L., Anderson, P.C. (1983) Somatic cell genetic systems in corn. In: Advances in gene technology: molecular genetics of plants and animals, pp. 147-157, Downey, K., Voellmy, R.W., Ahmad, F., Schultz, J., eds., Academic Press, New York.

Hibberd, K.A., Green, C.E. (1982) Inheritance and expression of lysine plus threonine resistance selected in maize tissue culture. Proc. Natl. Acad. Sci. USA 79, 559-563.

Lu, C., Vasil, I.K., Ozias-Akins, P. (1982) Somatic embryogenesis in _Zea mays_ L. Theor. Appl. Genet. 62, 109-112.

Lu, C., Vasil, V., Vasil, I.K. (1983) Improved efficiency of somatic embryogenesis and plant regeneration in tissue cultures of maize (_Zea mays_ L.). Theor. Appl. Genet. 66, 285-289.

CROWN GALL INDUCTION BY WILD TYPE AGROBACTERIUM TUMEFACIENS

This experiment is to examine the time course of crown gall formation on bean leaves afer inoculation by bacteria. It can be modified to examine infectivity differences between different strains of bacteria, effect of inoculum density, etc.

Bacterial culture. Use a 48-hr shake culture of bacteria grown in a medium of 0.8% nutrient broth, 0.1% yeast extract, and 0.5% sucrose at 27°C.

Plants. Use bean (Phaseolus vulgaris) about 6-7 days old. Plants are optimal when the primary leaves are 8-10 sq. cm in area.

Procedure. Dust the primary leaf of each plant lightly with 400 mesh carborundum and about 0.1 ml of the bacterial suspension. Take a Pasteur pipet and spread the mixture over the whole upper surface of the leaf while supporting the leaf with the fingers of one hand. The purpose of the carborundum is to abrade the leaf surface and to reate wounds necessary for the entry of bacteria. When the leaf is dry wash off the carborundum. Examine the leaves under a dissecting microscope that has illumination from below so that light is transmitted through the leaf. Tumors will first show up as dark regions in the lighter leaf background. After about 7-10 days they should be visible without microscopy.

Alternate procedure. Leaf tumors may be limited in growth. Those on the stem usually develop to larger size. Place a drop of the bacterial suspension on the side of the hypocotyl of 6-7 day old bean plants and make several punctures of the tissue through the inoculum drop, using a needle or pin. Observe tumor formation and growth in the next 2 weeks.

TOMATO GRAFTING AND SHOOT REGENERATION FROM WOUNDS

This experiment is designed to determine requirements for regeneration of de novo shoots from stem tissue of tomato, then to graft together stems of different genotypes and from the graft region regenerate new shoots that have the possibility of including cells of both genotypes in their meristems and thus to produce graft chimeras.

Regeneration. Use seedling tomato plants about 8 weeks old. Make a long slanting cut in an internode that has just completed elongation, using a single edge razor blade. To make reproducible cuts draw a vertical line on a piece of paper and another line that intersects the vertical line at about 20°. Lay the tomato plant with its stem along the vertical line and the internode to be cut over the intersection of the lines. Align the razor blade with the diagonal line and make the cut. Coat the cut surface of the stem with a layer of lanolin containing 0, 0.05, 0.1, or 0.5% parachloro phenoxy acetic acid, an auxin analogue. Make observations on the development of the cut surface at 2-3 day intervals, using low power of a dissecting microscope. New shoots should be visible in 7-10 days. It is important to remove axillary buds that begin to develop during the course of the experiment.

Grafting. Three kinds of grafts can be made: Slant grafts, in which the same kind of cut is made as used in the regeneration experiment; cleft grafts, in which the stem is cut only half way through at the same angle as used in making the sloping cut, then a second cut is made from the opposite side of the stem at the same angle so as to leave a deep V-shaped cleft in the stem; approach grafts, in which a slice of tissue is removed from the side of the stem. For all three kinds of grafts a second plant is prepared in the same way and the two partners are bound together so that the cut surfaces are in tight contact. Suitable binding materials are strips of Parafilm, Scotch tape, or strips of polyethylene sheeting (lunch bags are a good source) held together by alligator clips. It is important that the cut surfaces be in very close contact so that new cell growth from each partner of the graft will result in the formation of a graft union and so that efficient water and nutrient transport can occur from stock (the lower partner) to the scion (the upper partner). After about a week the graft should have formed. Remove the binding strip from each graft and slice through the stem of the scion, or in the case of the approach graft one of the partners, so as to leave only a thin layer, less than 1 mm thick, attached to the other partner of the graft. Coat this with lanolin (or the concentration of PCA found to be optimal from the regeneration experiment above) and observe in the following week or so for regeneration. In the approach graft it is necessary to decapitate the stock plant, and in all cases axillary buds should be removed as they start to develop.

Grafts can be made between partners with different leaf shapes, such as tomato with compound leaves, and pepper or eggplant which have simple leaves. Chimeric regenerated shoots can then be recognized as those in which the leaves differ in shape from each parent.

PROTOPLAST ISOLATION FROM NICOTIANA TABACUM XANTHI LEAVES

3 liters sterile water   10% Bleach

Enzymes: 1% Cellulysin, 0.1% Pectolyase in no hormone medium (NH), filter
          sterile.

1. Maintain shoot cultures of diploid or haploid Nicotiana tabacum MSO medium
    (Murashige-Skoog medium with no hormones). Grow in lighted chamber at 27°C
    with subcultures every 4 weeks.

2. Pot healthy rooted plantlets in soil, and grow in growth chamber with 16/8
    light/dark cycle and 23°C. Initially keep the humidity high by covering
    the plants with a plastic top. After several days this can be removed by
    gradually increasing the time spent without the cover. Fertilize the
    plants once per week.

3. After about 2 weeks, pick 5 to 6 leaves. Sterilize by rinsing for 5 min in
    1 liter of 10% bleach in a casserole dish. Suction off the bleach and
    rinse 3 times with 1 liter of sterile distilled water. Hang the leaves to
    dessicate.

4. Allow leaves to dry for about 1 hr. Filter sterilize NH just prior to use.
     Peel the leaves and float in NH in a petri plate. When done peeling the
    last leaf, remove the NH (can be refiltered and reused). Add enzyme and
    digest about 2 hr.

5. Gently pipet off the enzyme (it is possible to freeze the enzyme, and then
    reuse it several times by thawing and refilter sterilizing). Add 10 ml of
    NH to petri plate with leaves. Using a pipet shake the leaves to dislodge
    the protoplasts.

6. Pipet 5 ml into 15 ml tubes. Spin 80g for 15 min. Remove tubes and let
    sit 10 min. With a 1 ml pipet remove the top dark green layer of
    protoplasts. Pipet into a new 15 ml tube, add NH to 5 ml and repeat spin.
    Remove protoplasts and estimate density.

7. Dilute protoplasts with NH to a density of about 5×10^4 per ml. Pipet 5
    ml per tissue culture flask. Incubate in the dark for 2 days, then add 1
    ml of 5XH, gently mix, and place in low light.

8. After division starts, dilute with an equal volume of HHLS, and split into
    two flasks. When the colonies are just visible, mix with an equal volume
    of medium containing lukewarm 1% low-melting-temperature agarose. Let
    solidify, and continue incubation in low light.

NOTE: Leaf protoplasts require hormones for growth, however, we have found that
      they are healthier if isolated in NH, and then hormones are added a day or
      2 later.

28

PROTOPLAST FUSION - IN A TUBE

This is a bulk protocol. Many investigators have also been successful using microscale procedures.

1. Isolate protoplasts from <u>Nicotiana tabacum</u> Xanthi leaves as before. Isolate protoplasts from the <u>ws</u> mutant (white seedling) by slicing the leaves into thin strips, and then placing them in the same enzyme solution. When the leaves are digested, gently break up any leaf remains, then pipet the enzyme solution into a centrifuge tube. Do not discard the enzyme solution. Spin at 80g for 15 min, then collect the white protoplast layer at the top. Dilute into NH as in the green leaf protocol, and then float again.

2. Mix some of the two protoplast solutions (1 ml final). Ideally, the final concentration should be about 1×10^6, and should have equal numbers of each genotype. It is possible to mix the protoplasts of the two types together before the final centrifugation, and then spin them up together. Add an equal volume of 50% PEG (1 ml), gently mix. Incubate for 5 to 15 min. Add an equal volume of Ca-glycine 10.5 buffer (2 ml), gently mix and incubate for 10 min.

3. Add 10 ml of 0.6M mannitol. Gently mix. Spin 80g x 20 min. Resuspend pellet, very gently, in 5 ml NH. Pipet into flask and culture as before.

<u>50% PEG</u>
50% PEG 4000
50 mM $CaCl_2$
50 mM Glycine
0.1 M Sucrose
pH 5.5

<u>Ca-glycine 10.5</u>
50 mM $CaCl_2$
50 mM Glycine
0.4 M Sucrose
pH 10.5

<u>0.6 M Mannitol</u>

NOTE: The easiest way to make the PEG solution is to first make up the rest of the stock at double the final concentration. The PEG is then autoclaved as a powder in a glass bottle. The autoclave will melt the PEG into a sterile liquid; the solution can then be finished by mixing concentrated stock and PEG.

MAIZE GENETICS LABORATORY

I. Laboratory Outline
    A. Planting Considerations
        1. Selection of seed materials
        2. Genetic records system
        3. Field site selection
        4. Soil preparation
        5. Planting and germination of seed
    B. Plant Characteristics
        1. Anthocyanins
        2. Morphological markers
        3. Biochemical and physiological markers
        4. Recording plant characteristics
    C. Flowering Characteristics
        1. Female and male infloresence
        2. Pollination techniques
        3. Data management
        4. Harvesting and storage
    D. Kernel Genetics
        1. Aleurone markers
        2. Pericarp markers
        3. Endosperm markers
        4. Embryo and scutellum characteristics
        5. Cob and glumes
        6. Recording and managing kernel data
    E. Biochemical genetics
        1. Pollen mutagenesis (discussion)
        2. Staining and selection of pollen grains
        3. Cloning genes with transposable elements (discussion)
        4. Maize DNA minipreps
        5. Genomic cloning and Southern blotting with maize DNA
    F. Cultural Considerations

Genetic Records

Several systems of record keeping have been designed in order to maintain proper pedigree information. Essentially, the requirements are that every individual be identified by source, phenotype and genotype so that one can follow an individual's history and access important information readily. The following system was designed by Dr. Barbara McClintock and has been adapted by us. A computerized version will soon be available for release. The record system is essentially divided into three separate files:

1. <u>CULTURE FILES</u> These records are stored on 4 X 5" index cards and contain information regarding the source of the material to be planted in a particular year. It serves as a portable source of information during the planting period and during the initial growth of the plants and as a future permanent record of pedigree information. Each record is identified by a CULTURE NUMBER (1) which becomes the family number for the individuals planted from the ear. Both female and male family identification numbers (2) and genotypes (3) are given. Also included are the family subdivisions (4) assigned to each phenotypic class of kernels selected for planting purposes and a description of the kernel (5) and the number of kernels planted or germinated (6). Finally, the data are added regarding the number of seeds which germinate (7) and any unusual situations are noted (8) on the culture cards during the initial growing period.

2. <u>FIELD RECORDS</u> Once the seedlings are established in the field notebook is made which will store the information on the individual plant characteristics and genetic crosses to be made with each plant. This notebook is started with the information already collected in the culture records. Each culture is recorded (1), the parents (2), their genotypes (3), the family subdivisions (4) along with the phenotypic descriptions (5). Every plant in the field is assigned a consecutive PLANT NUMBER which will later serve to identify the parents of a particular ear (e.g. $7995A_1$ X $8004B_{10}$). As the plants reach maturity the somatic characteristics are recorded for each plant (6). Pollinations fall into three categories: crosses TO (male parent), BY (female parent) and selfed (X). Finally based on examination of the individual's progeny in the appropriate crosses, the genotypic information is subsequently recorded for each plant.

3. <u>EAR RECORDS</u> Analysis of the segregation data obtained from a particular cross usually requires that the ear be shelled and each phenotypic class counted and recorded. This data is transferred to a record consisting of the following information: the culture numbers both parents (1), their respective genotypes (2), the cob (3) features, the soft (4) and hard (5) glume characteristics. Include a schematic representation of any sectorial or positional information (6) about the kernels on the ear. Since not all crosses will be examined in such detail, we indicate in the field records which ears have been examined by a checkmark next to the cross which represents that a ear record has been initiated.

Data is further analyzed by comparing a series of crosses for similar information such as recombination frequencies between two or three selected marker. This information is best processed using Tables which include the segregation data from many crosses between similar plants.

Notes on Cultural Practices and Field Preparation for Maize

Useful References: Larson, W.E. and Hanway, J.J. Corn production. In: Corn and Corn Improvement. ed. Sprague, G.F. Am. Society of Agronomy. Chapter 11, pp 625-669.

A sucessful "crop" of maize results from selecting the best cultural practices, choice of seed and planting procedures, and method of pest control. The molecular biologist may find that a few plants grown in the greenhouse adequate for most experiments. However, for mutational analysis and linkage studies it becomes necessary to resort to field cultivation of relatively large numbers of plants. Selection of the optimum management controls must be evaluated each season and application of treatments at the proper time is critical. This outline is designed to give the reader a list of technical information which should be considered when planning field experiments with maize. It is designed only as an general backround which must be supplemented by consulting your local Agricultural Experiment Station's Cooperative Extension Service for information regarding situations which may be unique to your area (e.g. pathogens, planting dates, soil qualilty). The information we present is based on our experience of field experimental in the Northeastern region of the U.S.

1. Large Equipment: The following list of equipment is needed for proper field preparation. This is standard equipment for most agricultural schools. If no equipment is available, the entire field can be prepared and maintained with small 18 hp tractor with the appropriate attachments:

Tractor with 3 point hitch and PTO drive
Primary tillage system (e.g. moldboard plow)
Secondary tillage system (e.g. disk harrow or tractor-mounted rotortiller)
Seed bed preparation (e.g. meeker harrow)
Fertilizer and lime spreader
50 gal sprayer with boom attachment

2. Growth requirments.

   a. Soil: Ideally, a deep, medium-textured soil with a pH 6.0 would be ideal for plant growth. However, maize can be grown on widely different soils throughout the world. Plants grown in acidic soils are influenced by toxic concentrations of aluminum and manganese, and deficient concentrations of calcium and magnesium. In addition, soil pH will influence the availability of virtually all the major macro- and micro-nutrients and microbial interactions. It is essential to obtain a complete soil test report when growing maize on a field for the first time. This test, performed by Cooperative Extension Services, will include a report on the concentration of major plant nutrients, micronutrients, soil pH, percent organic matter, and soil texture. If you inform the Extension service of the crop to be grown, they will issue the appropriate fertility recommendations such as limestone and fertilizer application rates in lbs/acre. Limestone should be applied in the preceding fall in order to have time to influence soil pH by spring planting time. Fertilizer is applied after plowing and before disking to incorporate into the soil. A second application of urea is applied as a side dressing to plants approximately weeks after planting

   b. Moisture: Maize has a water requirment totaling 40 to 60 cm of evapotranspiration which varies considerably with the available soil moisture and enviromental conditions. A single stalk on a hot, windy day may use up to 2 liters of water. Since the water needs of maize during the summer in most areas will exceed the amount of rainfall, an irrigation system will probably be essential. The period of plant growth between tasseling to silking, usually during the peak of summer, is particularly critical. Any stress, such as indadequate water or fertility may delay silking for 2 or more weeks and cause reduce seed set due to inadequate supply of viable pollen at the time of silking. To avoid severe reduction in seed set, it is necessary to maintain soil moisture at adequate levels by irrigation. An irrigation system may be as simple as garden hose for a small field plot (¼l arce) but as the size of the planting area increases, so will the demand for a more elaborate, high volume irrigation system. The soil moisture can easily be monitered using a soil hygrometer. When moisture values fall below irrigation should be applied.

   c. Pathogens: The awareness and recognition of maize pathogens is essential when growing plants both in greenhouse and field situations. Different geographical areas are prone to different pathogens. Consult your local Extension Service for information on corn pathogens in your area. The inherent resistance or susceptibility of the plant may be the determining factor in the occurrence of disease. Inbred lines exhibit marked differences in resistance to certain pathogen. Stressed plants are more susceptible to disease and certain cultural practices can minimize disease occurrence. For instance, maintaining well-balanced soil fertility, soil moisture levels, and plot rotation can reduce plant infections. Unfortunately, proper protection, especially once an outbreak is identified, demands chemical application. Failure to control a disease outbreak will certainly mean lost experiments. A description of the major pathogens of maize can be found in Chapters 8 and 9, Corn and Corn Improvement (ed. G.F. Sprague, American Society of Agronomy Press)

Pollen Mutagenesis

Most effective method: Alkylating agent EMS
Best method: Parrafin oil technique

  Ref: M.G. Neuffer, 1978, pp. 579-600, Genetics and Breeding
        of Maize, ed. Walden, D.B.
      M.G. Neuffer and E.H. Coe, 1977, Maydica 22:21-28.
      M.G. Neuffer. MGCN 56:42
      M.G. Neuffer. Chapter 7. Maize for Biological Research.

Procedure:

1. Mix one part EMS stock solution with 15 parts light paraffin oil.

2. Mix one volume _fresh_ pollen with at least 15 parts treatment solution in a
tightly capped vial.

3. Close vial and shake periodically to prevent pollen from settling out.

4. After 45 min apply the pollen-oil mixture to fresh silks with a #10 camel
hair brush with vigorous stirring of the mixture between each application.

EMS Stock Solution: mix 1% EMS (Eastman #7830) in light paraffin oil (Fisher
#722268). Stir vigorously for 1 hr.

Pollen Staining for Waxy Classification

1. Dilute I_2-KI stock solution with 3 parts water.

2. Tease out anther in a drop of dilute I_2-KI on a slide.
   Examine under dissecting scope with light source from above.

3. Cut anther in two and push pollen out. Remove anther and place
   a coverslip over the pollen. Examine under lowest magnification
   using a microscope or under high magnification using a
   dissecting scope.

4. The Wx starch stains blue while the wx starch stains deep red.

I_2-KI Stock Solution:

      0.3 gm I_2
      1.0 gm KI
      100 ml distilled water

      Dilute with 3 parts water for pollen and endosperm
         classification.

MAIZE DNA MINIPREP

Stephen L. Dellaporta, Jonathan Wood, James B. Hicks

The topic of this protocol is rapid microscale methods for isolation of plant DNA without the use of ultracentrifugation with CsCl. The DNA produced is of moderately high molecular weight and serves as a satisfactory substrate for most restriction endonucleases and is suitable for genomic blot analysis. In addition to the rapidity and convenience of minipreps which permit a large number of samples to be processed in just a few hours, the small amount of tissue required (less than 1.0 grams) allows for molecular analysis of plants at a very young stage. Miniprep DNA yields from leaf tissue of most species tested to date are typically 50-100 ug per gram tissue, greater than 50 kb, and remarkably uniform from sample to sample.

The first miniprep procedure we reported for maize DNA isolation (Dellaporta et al, Maize Genetics Cooperation Newsletter, 1983) was adapted from a procedure commonly used for yeast DNA preparation (Davis et al., 1980). Since this report, numerous personal communications have demonstrated that the miniprep procedure or a modification thereof, can be applied to most plant species tested. For example, the method has been sucessfully used on <u>Nicotiana</u> <u>tabacum</u>, <u>N. plumbaginifolium</u>, <u>N. sylvestris</u>, <u>Lyscopericum</u> <u>sp.</u>, <u>Amaranthus</u> <u>sp.</u>, <u>Glycine</u> <u>max</u>, <u>Petunia</u> <u>hybrida</u>. Several modifications have been applied by these investigators and in our own laboratory in order to extend the application of miniprep procedures to other plant species. The selection of a particular protocol depends to a large degree on the plant species used. However, the procedure reported here was selected to be suitable for most situations.

Procedure

1. Weigh 1 gm of leaf tissue, quick freeze in liquid nitrogen and grind to a fine powder in a 3" mortar and pestle. Transfer powder with liquid nitrogen into a 30 ml Oak Ridge tube.

    It is imperative not to let the tissue thaw once frozen until buffer is added and not to cap the tubes while nitrogen is evaporating.

2. Add 15 ml of Extraction Buffer (EB): 100 mM Tris pH 8, 50 mM EDTA pH 8, 500 mM NaCl, 10 mM mercaptoethanol.

    For maximum DNA yields, the cells are further broken by grinding the mixture at a low setting (about 3) with a Polytron (Brinkmann Instruments, Inc.). However, this step is optional.

3. Add 1.0 ml of 20% SDS, mix thoroughly by vigorous shaking, and incubate tubes at 65°C for 10 min.

4. Add 5.0 ml 5 M potassium acetate. Shake tube vigorously and incubate 0o for 20 min.

    Most proteins and polysaccharides are removed as a complex with the insoluble potassium dodecyl sulfate precipitate.

5. Spin tubes at 25,000 x g for 20 min. Pour supernatant through a Miracloth filter (Calbiochem) into a clean 30 ml tube containing 10 ml isopropanol. Mix and incubate tubes at -20° for 30 min.

6. Pellet DNA at 20,000 x g for 15 min. Gently pour off supernatant and lightly dry pellets by inverting the tubes on paper towels for 10 min.

7. Redissolve DNA pellets with 0.7 ml of 50 mM Tris, 10 mM EDTA, pH 8. Transfer the solution to an Eppendorf tube. Spin the tubes in a microfuge for 10 min to remove insoluble debris.

8. Transfer the supernatant to a new Eppendorf tube and add 75 ul 3M sodium acetate and 500 ul isopropanol. Mix well and pellet the clot of DNA for 30 sec in a microfuge. Wash pellet with 80% ethanol, dry, and redissolve in 100 ul 10 mM Tris, 1 mM EDTA, pH 8.

Precipitation from 0.3 M sodium acetate using relatively small amounts of isopropanol (about 0.6 volumes) has been reported to separate high molecular weight DNA from polysaccharides (Marmur, 1961). The sodium acetate also yields a tight fibrous precipitate that is easily washed and dried. The DNA will dissolve readily if allowed to rehydrate at 4°C for 1 hr followed by light vortexing.

Minipreps can be stored for several months without evidence of degradation and can be cut with a variety of restriction enzymes and ligated without further purification. We find that 10.0 ul of miniprep DNA is sufficient for a single 8 mm lane in an agarose gel which is to be used for filter hybridization with single-copy probes. Heat-treated RNAase must be added to the restriction reaction to digest contaminating RNA in each prep. Hence, a typical reaction would contain the following:

Miniprep DNA	10.0 ul
10X Restriction Buffer	3.0 ul
0.5 mg/ml RNAase	2.0 ul
Eco RI	8 units
dH20 to 30 ul.	

Digestion is usually complete after 3 hours at 37°C. Occasionally, minipreps are difficult to digest with certain enzymes. This problem can be overcome by adding 5.0 ul of 0.1 M spermidine to the entire miniprep before digestion (see Focus 4(3):12, 1982).

REFERENCES

1. Davis, R.W., Thomas, M., Cameron, J., St. John, T.P., Scherer, S., and Padgett, R.A. Rapid DNA isolation for enymatic and hybridization analysis. Methods in Enzymology 65: 404-411.

2. Marmur, J. 1961. A procedure for the isolation of deoxyribonucleic acid from micro-organisms. J. Mol. Biol. 3:208-218.

MAIZE ADH ISOZYMES - STARCH GEL ELECTROPHORESIS

A. Preparation of Gel

1. Tape gel molds together, six for each set, making certain that bottoms of inside molds are even. Apply mineral oil to insides of molds and to one side of each top.

2. Weigh out 33g of starch and place in dry 3 l boiling flask.

3. Measure 300 ml of gel buffer (see below) in graduated cylinder and pour into boiling flask. Swirl immediately to avoid formation of clumps.

4. Light Bunsen burner.

5. Using asbestos glove, swirl flask over heat at top of light blue flame until gel forms. At first liquid mixture will be easy to swirl. Then it will thicken and be harder to swirl. Finally when it becomes a gel it will be again easy to swirl and small bubbles will form on the side.

6. Evacuate flask with aspirator. At first gel will boil and form frothy small bubbles. Then large bubbles will appear and finally no bubbles at all. Remove vacuum hose after large bubbles form but before all boiling ceases.

7. Pour gel into molds and let stand for 5 min. Wash boiling flask immediately with hot water.

8. Place tops on gel molds carefully being certain not to trap air bubbles. Press gently to remove excess gel.

9. Place in refrigerator for at least 20 min.

10. Remove tops and excess starch.

11. Make cuts in gel with narrow razor blade.

B. Preparation of Sample

1. Dry Seed Scutellar Slice

   a. Remove small sliver of scutellum with razor blade and place in numbered well of plastic block. Take care not to injure embryo.

   b. Save seeds on index card covered with masking tape, sticky side up. Write date, seed source and seed numbers on card in ink.

   c. Place 60 ul of diluted sample buffer in each well and macerate sliver with flattened glass rod. Rinse rod thoroughly in distilled water between each sample.

38

d. Place Whatman 3 MM filter paper wick (6 mm x 8 mm) in each well.

e. Blot with and place in slot on gel (3 abreast in each slot = 54 samples per run).

f. Place Saran Wrap on top of gel.

2. Sixteen Day Old Endosperm

a. Harvest seed 16 days post-pollination, place in Ziploc bag and store in freezer.

b. Separate endosperm from embryo and place in well of plastic block.

c. Squash with glass rod.

d. Place filter paper wick in well and allow liquid to absorb.

e. Remove wick, blot and place in appropriate slot of starch gel.

f. Place Saran Wrap on top of gel.

C. Running Gel

1. Remove gel from chamber and place on lab bench.

2. Slice gels in shallow molds and place bottom sides in tray.

3. Pour stain over gels.

a. Activity stain--leave until desired intensity develops and then replace with water.

b. General Protein Stain (Coomassie Blue)
Stain for 1/2 hr., pour off stain and then distain with 10% acetic acid, 50% MeOH.

E. Recipes

Gel Buffer (0.03 M Na Borate, pH 8.5)
per liter

1.85g H_3BO_3
0.21g NaOH

Chamber Buffer (0.3 M Na Borate, pH 8.5)
per liter

18.5 g H_3BO_3
3.4 g NaOH

10X ADH Extraction Buffer

5 ml 1 M Tris-HCl pH 8.0

39

231.45 mg dithiothreitol
to 50 ml with glass distilled H_2O

ADH Stain

    10   ml 1m Tris-HCl pH 8.0
     2   ml 0.01 M NAD
     1   ml 0.01 M Nitroblue tetrazolium
     0.5   ml 95% EtOH
    to 100 ml with glass distilled H_2O

Just before pouring over gels, add 1 ml of 0.01 M phenazine
methosulfate.

General Protein Stain

    1 g Coomassie Blue/liter of 10% acetic acid.

TOBACCO CELL CULTURE RNA -- GUANIDINE THIOCYANATE

Care is needed to prevent contamination of materials with RNase. Materials should be autoclaved, or washed with NaOH, and some stock solutions can be treated with DEPC, and then autoclaved. (DEPC is not very effective in Tris buffers). Use sterile plastic pipets and tubes. Wear gloves at all times.

Lysis Buffer (make up just before use)   5.7 M CsCl

5M Guanidine Thiocyanate (Eastman Kodak)   20% Sarkosyl
50 mM Hepes pH 7.4
Heat stir for 30 min. H_2O
Pour into top of filter sterilizing unit.
Add mercaptoethanol to 5%, filter sterilize.   3 M Na acetate
Heat 65° for 30 minutes.

1. Collect plant cells from culture on Whatman filter paper with vacuum. Weigh cells. Add lysis buffer to cells at 4 ml per gram, homogenize with Polytron on high speed until well emulsified (usually 2 or 3 - 15 sec bursts)

2. Spin 12,000g for 20 min, save supernatant. Add 1/9 volume of 20% sarkosyl and 1/9 volume of 5.7 M CsCl. Heat to 65° for 10 min.

3. In tubes for SW27 rotor, or any equivalent large volume swinging bucket rotor, fill tubes to 2/3 full with RNA supernatant. Underlay with 1/3 volume of 5.7 M CsCl. Spin for 16 hr at 25,000 rpm, 20°C.

4. Resuspend in 1 ml of H_2O. Heat at 65°C to get into solution. Spin 4000 rpm for 15 min to remove insoluble materials. [Options: If you feel your RNA needs further work, you can phenol extract, then chloroform extract, and then ethanol precipitate]
   Add 0.2 volumes of Na acetate, then 4 volumes of ethanol, and precipitate at -20° (overnight) or -70° (1 hr). [For in vitro translations, precipitate with K acetate].

5. Resuspend in 1 ml of H_2O. Read the 260/280 absorbances.

NOTES: (1) This is total RNA, suitable for blotting or in vitro translations. (2) It takes about 200 ml of log phase cultures, or 50 ml of stationary phase cultures, to get the proper amount of material for 1 tube in the rotor. (3) The length of time of the spin, and the relative amount of the CsCl cushion are important variables in removing starch, DNA, and protein. If you have trouble adapting this protocol to another tissue or organism, you should try varying the cushion and length of spin.

MAIZE PROTOPLAST ISOLATION

NOTE: This technique yields protoplasts from MGI's strain of Black Mexican
        Sweet (BM7) which will regenerate into callus culture at a rate of 0.1%.
        Other techniques using less expensive enzymes will efficiently produce
        maize protoplasts, but attempts to regenerate these or other BM7
        protoplasts into callus have been unsuccesful. It should also be noted
        that this technique was developed for producing protoplasts from BM7
        suspension cells and in no way should be regarded as a suggestion that
        it might possibly work for any other source of maize cells. Such is the
        state of the art for maize protoplasts.

        The technique presented below was developed by Paul Anderson at
        Molecular Genetics, Inc., Minnetonka, MN, (after Chourey) and has been
        slightly modified by Steve Ludwig and Robert Pohlman.

MATERIALS

    5-8 gm of fresh BM7 suspension cells. Prepared as follows: inoculate 200
ml of MS + 2% sucrose, 2 mg/ml 2,4-D (in a 500 ml flask) with 20 ml of late log
- early stationary BM7 culture; incubate at 26-30°C / c. 250 rpm for 3 days. 4
to 6 gm can routinely be obtained from one 3-day old 220 ml culture.

    Cellulase (only from Worthington, #'s LS0002601, 03, or 04)
    Pectinase (only from worthington, #'s LS0004297, 98, or 99)
    D-Mannitol
    CaC12
    50 ml screw cap sterile plastic "Falcon"-type tubes
    46 micron filters (Bellco supplies stainless steel ones)
    Clinical centrifuge
    MS+sucrose, 2,4-D
    Protoplast Regeneration Medium (PRM) (modified Kao's)
    Hemocytometer
    Filtering apparatus for sterilizing enzyme solution
    Rotary shaker (50 rpm)
    Some sort of Pastuer pipet/aspirator-type apparatus (optional)
    Low Melting Point Agarose (BRL) (optional)
    35 x 15 mm petri plates (optional)
    Calcofluor White (optinal)
    Fluorescein diacetate (optional)
    Fluorescent microscope (optional)

STOCK SOLUTIONS

    Enzyme solution:
        1% Cellulase (if wealthy, use 2%)
        0.25% Pectinase
        0.2 M Mannitol
        80 mM CaC12

    Adjust unbuffered to pH 6.0, filter through 0.2u; best to prepare just
prior to use but can be stored at -70 for 3 weeks.

    MS + sucrose, 2,4-D:

PRM (modified Kao's):

    (if replating protoplasts, also prepare 100 ml of 2xPRM)

Calcofluor White:
    0.1% in 0.2M mannitol (0.5-0.7M for tobacco)

Fluorescein diacetate:
    0.1% in 50% acetone/50% water

PROCEDURE

When in Doubt *** Always *** be *** Gentle *** With The Protoplasts

All transfers should be performed GENTLY down the walls of the vessel.

All steps are at room temperature unless otherwise noted.

1. Let two 220 ml cultures of BM7 suspension cells settle at room temperature for 10 min. Pull off most of the supernatant.

2. Transfer c. 50 ml of the remaining cell suspension to two 50 ml plastic cfg tubes and let settle until 5 ml (Ludwig's Law of Conversion: 1 ml = 1 gm) of cells have settled. Pull off excess supernatant and unsettled cells.

3. Add 9 volumes of enzyme solution (45 ml) to each batch of cells, gently mix and incubate at room temperature for 3 hr, at 50 rpm.

4. Gently filter preparation through a sterile 46 micron filter. This is the appropriate size for BM7 protoplasts; larger cell types, of course, would require a larger filter at this step.

5. Cfg at 75-100 rpm in clinical centrifuge for 5 min. Remove super. The more gentle, the better.

6. GENTLY wash twice with 30-50 ml PRM.

7. Count protoplasts with hemocytometer. Usually 5 gm yields about 10^7 protoplasts.

NOTE: At any time during protoplast isolation one can stain a small aliquot with Calcofluor White and observe the cells with a fluorescent microscope. Any cells still retaining cell wall material will glow white-blue. Also, viability can be checked by staining with fluorescein diacetate. Living cells will fluoresce.

REGENERATION OF CALLUS (optional)

a. Prepare 2.0% low melting point agarose in water by autoclaving. Mix with an equal volume of 2X PRM.

b. Resuspend protoplasts in PRM at a density c. 2×10^5/ml.

c. Combine 0.5 ml each of the agarose/PRM and resuspended protoplasts in a 35x10 mm petri plate.

d. Seal petri plates THOROUGHLY with parafilm and place in a "humidity box" (anything that will help keep the plates from drying out). Incubate at 26-30°C in the dark.

NOTE: One should be able to observe cell divisions at about two weeks. Before then, the best rule of thumb is to leave them alone. (If you must look, it's the ones that look dead that are in reality probably the survivors.)

REFERENCES

Paul Anderson, Molecular Genetics, Inc., Minnetonka, MN.

Steve Ludwig and Robert Pohlman, Jo Messing's Lab (612) 376-9273.

Chourey and Zurawski, Theor. Appl. Genet. vol. 59: 341-344 (1981).

Kao and Michayluk, Plants, vol 126: 105 (1975).

Kao and Michayluk, Z. Planzenphysiol. vol 96: 135 (1980).

LIPOSOME PREPARATION

Materials

    TES (SIGMA)
    Bovine L-phosphatidyl-L-Serine (SIGMA) (Assume mw = 850)
    cholesterol (Sigma)
    chloroform
    ether (diethyl)
    L-histidine
    NaCl
    EDTA
    D-Mannitol
    Calcium oxide   )
) for ether distillation
    $FeSO_4-6H_2O$)

    5 1/4" Pasteur pipets, bulbs
    1.5 ml Epp tubes
    15 ml Corex cfg tubes (sterile and very well rinsed)
    Tweezers
    Sonicator: microtip: Branson sonic Model S125
               bath: Model 1125PIG (Laboratory Supplies, Hicksville, NY)
    Roto-vap w/sleeves or 500 ml round bottom flasks
    Vortex mixer

Stock Solutions

    Test buffer: 2 mM TES pH 7.4
                  2 mM L-histidine
                150 mM NaCl
                0.1 mM EDTA

    Cholesterol: 5 umoles/ml $CHCl_3$ - Store -15°C
    PH-serine: 10 mg/ml MeOH/$CHCl_3$ (as supplied by Sigma) Store -15°C
    3X Ether: Distill 400 ml diethyl ether, 1g CaO, and 10g $FeSo_4-6H_2O$.

Warm to reflux (do <u>not</u> use an open flame). Also, an ice water cooled condenser is required. Discard the first 50 ml of distillate. Collect the next 300 ml in an iced flask. Repeat twice, collecting 200 ml, and then 100 ml. Store over 1/4" H_2O in a dark bottle at -15°C.

Ideally, ether should be redistilled before each use. However, 4-6 wks at the above storage conditions appears to be okay. I've also been told that one should filter serilize the ether with 0.2 um poly-carbonate filters (Bio-Rad) but I just have not been able to bring myself to do that.

IX Liposomes (Rev) (all steps at room temperature)

1. Mix: 42 ul (0.5 umoles) pH-serine
             100 ul (0.5 umoles) cholesterol in 15 ml corex tube.

2. Dry in roto-vap (700 mm Hg - 28").

3. Redisolve in 200-300 ul ether.

45

NOTE: It is important that after sonicating small preparations that the resulting suspension is still "organic phase" dominant (an aqueous suspension in an organic solvent). Since the microtip sonicator evaporates more ether than the water bath-type sonicators, this amount of ether is required. For bath sonication, 100-150 ul of ether is sufficient.

4.   a.   bath sonication - add 5-100 ug DNA in 45 lambda TES+ to lipids. Vortex 1' at setting #3. Sonicate 15-30".

     or

     b.   microtip sonication - place DNA/TES+ in Epp tube, add resuspended lipids. Sonicate 3-4 times x 0.5-0.75 sec/burst at setting #1. With a fresh Pasteur pipet, transfer back into a fresh 15 ml Corex tube.

5.   Apply vacuum (350 mm/13.7" Hg) at 50 rpm until a viscous gel forms. Briefly vortex gel (#3 - 15-45").

     NOTE: In this step we are reducing the "ether dominant" suspension towards an "aqueous dominant" suspension of liposomes. The gel phase is the transition point. Ideally, and as is easily obtained in larger (10x) preps, the gel is somewhat "brittle" and breaks into 2-3 mm pieces upon vortexing. This increased surface to volume ratio, of course, facilitates the further transition to the liposome suspension. With smaller preps the transition phase can be rather transitory and difficult to identify. Usually one has to settle for a very thick gel that smears nicely along the walls of the tube when vortexed, but does not break into pieces. This is acceptable, but may require further vortexing in Step 6.

6.   Apply vacuum (700-730 mm/27.5-28.7" Hg) at 50 rpm. If gel does not "melt" into suspension of lipsomes, revortex tube and reapply vacuum. continue vacuum until ether is completely removed. If gel "bumps," revortex.

7.   Add 100 ul TES+, gently mix. (optional: overlay with inert gas)

8.   Liposomes can be separated from unincorporated material by various methods (see below), if necessary. The simplest is to collect by centrifugation at 48K x g x 20'. Resuspend in appropriate buffer/medium (TES+ or 10 mM Tris, pH 7.0, 50 mM KCl, 0.4 M Mannitol, 1 mM EDTA).

Determination of Encapsulation Efficiency

From: Fraley Workshop on Liposomes

I. Molecular Sieve Chromatography

   Note: CF (from Sigma) is very acidic. You will have to readjust to pH
          of the dye solution.

   1. Prepare REV containing the fluorescent dye, carboxy fluorescein (CF, dye solution is 75 mM, E 1mM/493 = 70.

2. Adjust the final volume of the liposome preparation to 1 ml with Tris buffered saline (TBS, 10 mM Tris, 150 mM NaCl, pH 7.0) and load the preparation on a small Sephadex G-75 column (use 10 ml pipette for column).

3. Collect the liposome fraction (leading peak) in a volume of 3-4 ml -- record volume.

4. Assay CF by diluting sample with TBS buffer (10^2-10^4 fold containing 0.1% Triton X-100 to disrupt the vesicles (note color change as CF is released and quenching reduced) and reading absorbance at 493 nm.

5. Assume 80% recovery of the total phospholipid and calculate the encapsulation efficiency of the preparation as:

    a) % aqueous sample entrapped
    b) ul sample encapsulated/umol phospholipid

6. Applaud if you have 35-40% encapsulation.

II. Flotation on Discontinuous Polymer Gradients

1. Prepare REV.

2. Adjust the final volume of the liposome preparation to 0.4 ml with sample buffer (SB) and transfer to a sterile SW50.1 centrifuge tube (sterilize by washing with 70% ethanol).

3. Rinse sample tube with 0.2 ml of SB and transfer to centrifuge tube (repeat twice). Add 1.0 ml of SB & 30% ficoll to centrifuge tube. Mix gently by vortexing.

4. Gently layer 2.5-3.0 ml of SB + 10% ficoll over the sample solution. Finally, layer SB (0.5-1.0 ml) over the ficoll solution to within 1/8" of the top of the tube.

5. Centrifuge at 25-30K for 25.0 min @ 20°C and then collect the liposome band from the SB/10% ficoll interface with a pipette in a final volume of 1.0 ml. Note, liposomes may aggregate in the ficoll solutions but will disperse upon vortexing.

6. Assay encapsulation efficiency. Assume 80% recovery of phospholipid and calculate encapsulation efficiencies as in part I.

7. Store vesicle preparations at 4°C under inert atmosphere (N_2 or Ar) if possible.

References:

Papahadjopoulos, et al. (1981) Techniques in cellular physiology, P114:1-18. A good review.

Fraley, et al. (1982) Proc. Natl. Acad. Sci. 79:1859-1863.

_____ et al. (1980) J. Biol. Chem. <u>255</u>:10431-10435.

Szoka and Papahadjopoulos. (1978) Proc. Natl. Acad. Sci. <u>75</u>:41-94.

Fraley, et al. (1982) Proc. Natl. Acad. Sci. <u>79</u>:1859-1863.

LIPOSOME FUSION TO PROTOPLASTS

Materials

Liposomes
Protoplasts
Tris
D-Mannitol
$CaCl_2$
Poly (vinyl alcohol) (PNA) Poly Sciences #2975
  88% mole hydrolyzed mw = 25,000
50 ml Plastic cfg "Falcon"-type tubes w/screw cap (sterile)
60 x 20 mm Petri plates
Parafilm

Stock Solutions

TBM + PVA 5 mM Tris pH 7.0
  0.2M (maize) or 0.5M (tobacco) Mannitol
  0.5 mM $CaCl_2$
  10% PVA

Note: It is best to add Tris, $CaCl_2$, and PVA. Heat in autoclave to solubilize the PVA. Filter through WHATMAN, etc. filter to remove "fish" supplied with the PVA. Then add mannitol and then autoclave to sterilize.

TBM - as above, but without PVA.

Appropriate plant cell culture medium.

   1. Combine 10^7 protoplasts (for transient expression, otherwise 10^6) (in 10.5-1.0 ml) and a "1X" aliquot of liposomes (10-500 moles of phospholipid) in sterile 50 ml tube. Set 5' at 25°C.

   2. Add 9 volumes of TBM + PVA. Incubate at 25°C x 20'.

   3. Bring to 50 ml with TBM. Pellet protoplasts by cfg ("100 x g x 10').

   4. Wash once or twice with TBM, once with culture medium.

   5. Resuspend in 20 ml culture medium.

   6. Plate 10 ml/plate, seal w/parafilm. Set in dark at 26-30°C.

References:

Papahadjopoulos, et al. (1981) Techniques in cellular physiology, P114:1-18. A good review.

Fraley, et al. (1982) Proc. Natl. Acad. Sci. 79:1859-1863.

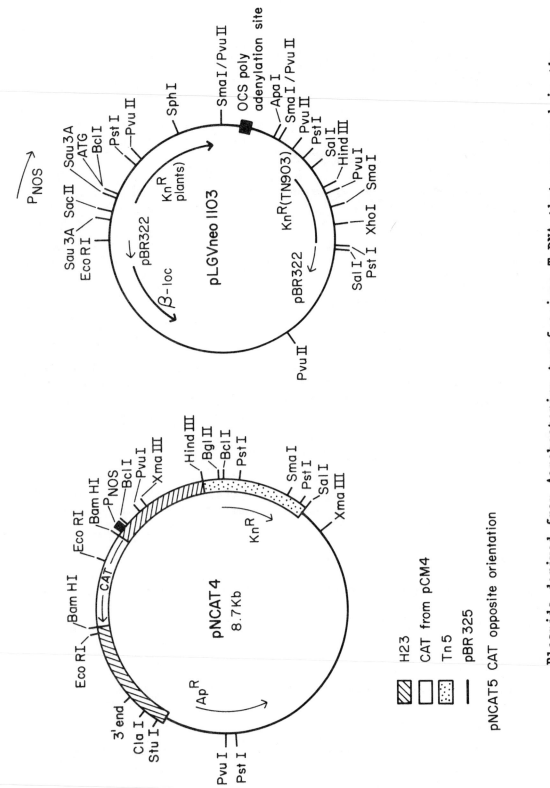

Plasmids derived from _Agrobacterium tumefasciens_ T-DNA that are used in the liposome and co-cultivation experiments, provided by Marc van Montagu (Rijksuniversiteit, Gent, Belgium). These are based on gene fusions of a procaryotic antibiotic resistance gene to the nopaline synthase promoter (p-nos) of T-DNA, which functions in plants. pNCAT4 has p-nos in front of chloramphenicol acetyl transferase, and p1103neo has the Tn5 kanamycin resistance linked to p-nos.

TOBACCO SUSPENSION CULTURE / AGROBACTERIUM CO-CULTIVATION

MS1 Liquid

MS1 Kan Plates
MS1 medium plus
50 ug/ml kanamycin
0.5 mg/ml cefotaxime

MS1 Top Agar
MS1 agar medium, aliquoted
remelt just prior to use
50 ug/ml kanamycin
0.5 mg/ml cefotaxime

Cefotaxime 10X
5 mg/ml
filter sterilize

Kanamycin 10X
0.5 mg/ml
filter sterilize

Minimal Bacterial Medium
Miller's A, or M9

NOTE: add antibiotics to medium after autoclaving, when it is lukewarm. For Top Agar, add the antibiotics just prior to use.

1.   In minimal medium, and under appropriate selection, grow an Agrobacterium strain containing the nopaline synthase Tn5 neomycin fusion plasmid overnight at 30°C.

2.   Have a finely divided suspension culture of Nicotiana in mid-log phase growing in MS1. Add 1 ml of Agrobacterium to 50 ml of suspension culture. Rotary incubate for 2 days at 27°C.

3.   If the culture appears thick with tobacco cells, split it 1:1 with fresh MS1 medium. Pour suspension cultures into a graduated 50 ml tube, and let settle for 20 min. Remove the supernatant, 5 mls of Cefotaxime-10X, and fill tube with fresh MS1 to 50 ml. Pour into a new flask and incubate for 3 days more.

4.   Have the MS1 Antibiotic plates ready at room temperature. Melt top agar, cool to lukewarm, and add antibiotics. Pour the suspension culture into a 50 ml graduated tube, and let settle. Remove supernatant, and add a volume of top agar equal to that of the cells remaining in the tube. Gently mix cells and agar, and then pipet 10 ml onto MS1 antibiotic petri plates. Let the plates sit while gelling, and then carefully wrap with parafilm. Incubate in the   dark   at   27°C.

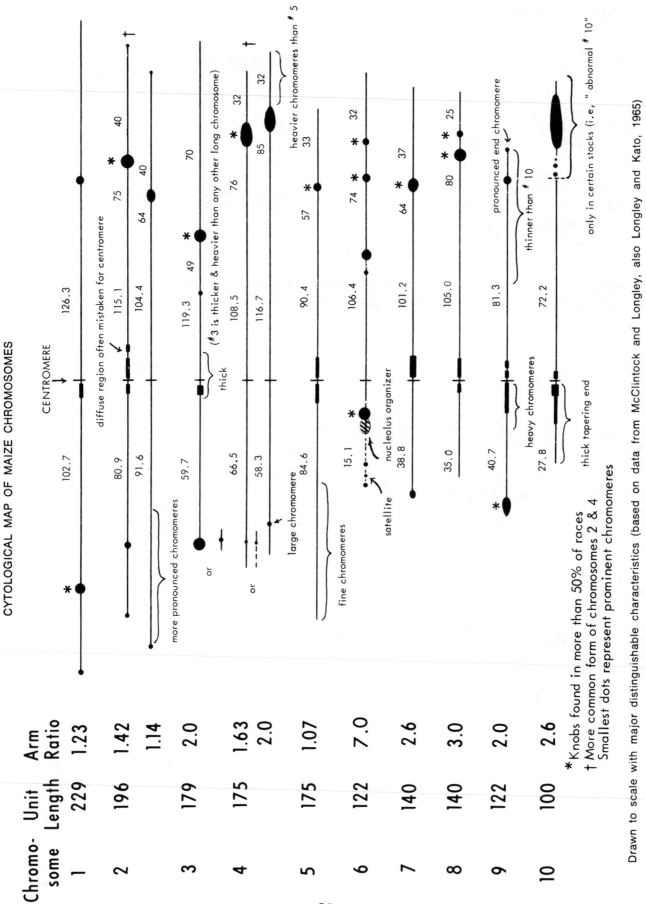

CYTOLOGICAL MAP OF MAIZE CHROMOSOMES

Drawn to scale with major distinguishable characteristics (based on data from McClintock and Longley, also Longley and Kato, 1965)

Reproduced from *Mutants of Maize*, 1968, p. 4, by Neuffer, Jones, and Zuber, by permission of the Crop Science Society of America.

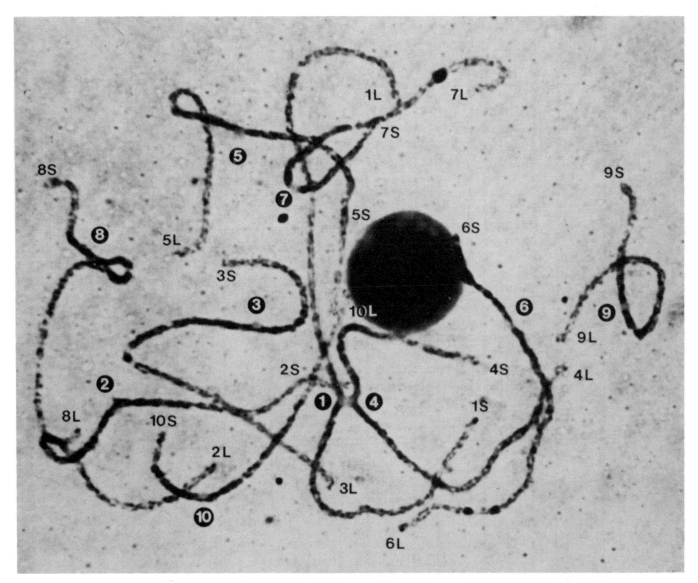

Maize Pachytene Chromosomes

Centromere locations are indicated by the adjacent solid circles with white numbers. The long and short arms of each chromosome are indicated by the chromosome's number and L and S respectively which are arranged near the terminal regions of the chromosome arms. The lengths of the arms and their ratios are listed below. Note that these do not agree in all cases with those shown with the cytological map, this is evident for chromosomes 1, 3, 4, 6 and 8. In the above figure arms 1L, 3L, 4S, and 8L are stretched somewhat and this results in the discrepancy of the ratios of these chromosomes. In measuring the length of arm 6S in the above figure the entire NOR and satellite region was included and this region was not included in the measurements used in preparing ratios listed with the cytological map (see preceding page). The pachytene spread was prepared and photographed by John T. Stout and the negative used to prepare the figure was kindly provided by Ronald L. Phillips. Reprinted from W.F. Sheridan, in Maize for Biological Research, pp. 37–52. Plant Molecular Biology Assn., Charlottesville, VA [1982].

Chromosome arm lengths in above figure (mm) and ratios

Chromosome	Long arm	Short arm	Ratio
1	135.20	92.53	1.46
2	101.33	70.60	1.43
3	112.23	48.30	2.32
4	89.23	62.63	1.42
5	82.70	72.80	1.14
6	88.73	32.60	2.72
7	92.33	35.50	2.60
8	96.87	29.73	3.26
9	72.53	36.13	2.01
10	74.66	27.66	2.70

MAIZE CYTOGENETICS

Chromosome analysis
Meiosis
Pachytene analysis
In situ hybridization
Mitosis — root-tips, cell suspension cultures

Changes in chromosome structure
Interchanges
Quartet analysis
Paracentric inversion

Changes in chromosome number
Monosomics
Tetraploid

Heterochromatin
B chromosomes
Abnormal chromosome 10
2NOR, Illinois Reverse High protein
Zea diploperennis x Zea mays

Note: Please also look at the section by Burnham in the Supplement

Meiosis in Maize

Microscopy
Life cycles
  1. Megasporogenesis
  2. Microsporogenesis
  3. Function of meiosis in life cycle

Meiotic stages
Slide preparation

OBJECTIVES

  -to learn technique of preparing slides of sporocytes
  -to identify the stages of meiosis in maize sporocytes

Read Burnham (Supplement on Cytogenetics) for a written description of
preparing and staining the sporophytes. Once you have made several slides of
different stages, ask yourself these questions:

  *What are the visible differences between Div. I and Div. II stages?
  *When does the nucleolus disappear and reappear?
  *How many nucleoli are there?
  *What do the microspores look like if the slide is heated too much?
  *Can any details of chromosome morphology be seen, such as knobs,
    constrictions, satellites and centromeres? Can you identify individual
    chromosomes by these features (see photos in Rhoades' article)?
  *Why are relative lengths used instead of actual lengths for identifying
    chromosomes?
  *Are these cytological features the same in A188 as in W23?
  *Are there any aberrations in your material, such as fragments of
    chromosomes, abnormal pairing or bridges?
  *At what stages would these aberrations be visible?
  *At pachynema, is there any pairing of knobs or centromeres from
    non-homologous chromosomes?
  *Can you tell when two chromosomes are overlapping versus forming a chiasma?
  *Is this easier to do using the oil immersion lens or the high dry objective?

Practice sealing a slide with wax. This will not preserve it permanently, but
it allows longer viewing without constantly adding more liquid which can change
the staining.

References:

Rhoades, M.M. 1950. Meiosis in maize. J. Heredity 41:59-67.

Burnham, C.R. 1962. Discussions in cytogenetics, Ch. 1.

Carlson, W.R. 1977. The cytogenetics of corn. In Corn and corn improvement.
(ed. G. Sprague) ASA Agronomy Series, Vol. 18.

Microscopy

55

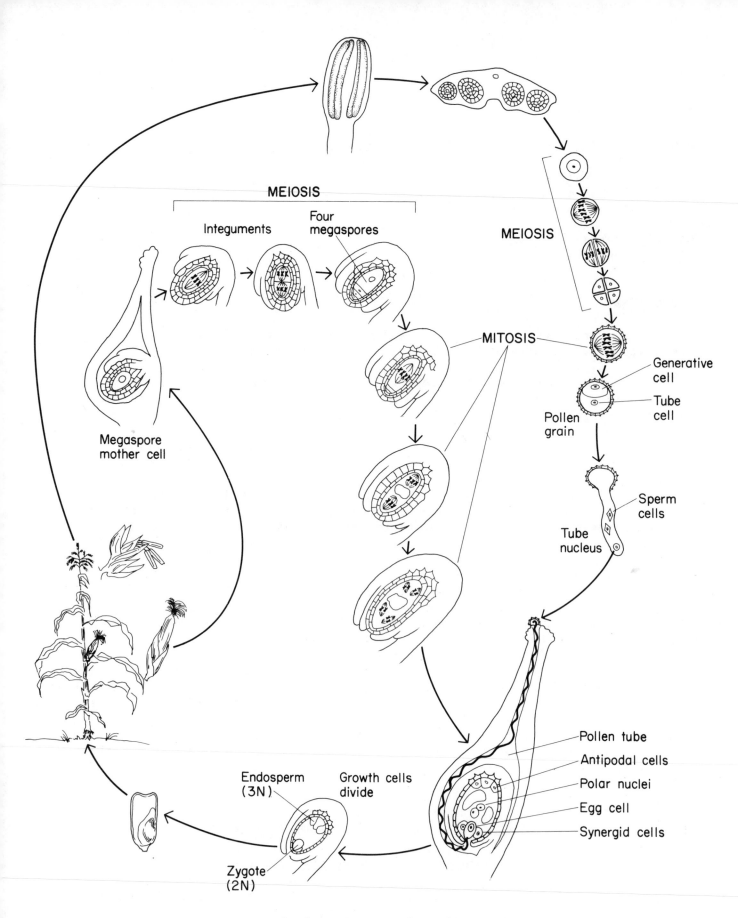

MEIOSIS

Integuments   Four megaspores

MEIOSIS

MITOSIS

Generative cell

Tube cell

Pollen grain

Megaspore mother cell

Sperm cells

Tube nucleus

Endosperm (3N)   Growth cells divide

Pollen tube

Antipodal cells

Polar nuclei

Egg cell

Synergid cells

Zygote (2N)

Meiosis and the maize life cycle

```
Basic Parts:  eyepiece
              body tube
              objectives (low and high power, oil immersion)
              stage
              coarse and fine focus knobs
              condenser
              iris diaphragm
              light source (with rheostat)
```

Magnification

In general, magnification is the size of the image relative to the original
size of the object. For the objective lens, this is calculated as the ratio of
the image size produced to the size of the object at a specified body tube
length (usually 160 mm). The magnification of the eyepiece is calculated on
the basis of a virtual image of 250 mm from the eye point of the ocular. Total
magnification is simply the product of these two figures.

Resolution

Resolution is the ability to distinguish and separate two adjacent objects.
The resolving power of a lens depends upon the angle through which it can
gather light. This angle is called the angular aperture and has a theoretical
maximum of 180 degrees. However, other factors also affect the resolution such
as the refractive indices of the lens, slides, cover glass, etc. Therefore,
another expression, the numerical aperture (N.A.) more nearly represents the
resolving ability:

$$N.A. = n \times sine\ 1/2\ A.A.$$

n being the refractive index of the least refractive material. The greater the
N.A. value, the greater the resolution. N.A. values range up to a practical
limit of about 1.4.

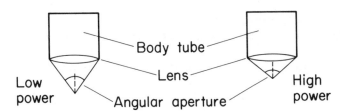

N.A. is dependent on the refractive index (n) of the least refractive medium in
the light path. By using an oil immersion lens system (n of oil = 1.52
compared to air n = 1.00) the resolution can be increased.

Depth of Field

The thickness of the object which seems to be in focus at any one time is
called the depth of field. The greater the resolving power, the less the depth
of field will be.

Koehler illumination

Koehler (pronounced "curler") illumination for microscopy is a system devised by August Koehler in 1893 which assures 1) an evenly lit field at the object plane and 2) the widest possible cone of light to fill the front objective lens (for optimal resolution). To achieve this it is necessary to 1) focus the filament of the illuminator onto the iris of the substage diaphragm and II) focus the image of the iris diaphragm of the illuminator onto the object, the steps to take are as follows:

1. Set the light source about 12" directly in front of the microscope.

2. Close down the light source iris to its smallest aperture and open the substage iris to its fullest aperture.

3. Using the knob at the side of the lamp, focus an image of the filament on a piece of paper held on the mirror. Position that image in the center of the mirror. This accomplishes condition I.

4. Focus the microscope on the object using that objective (oil immersion) that you will be using for critical work. Then rack the substage condenser up to almost the top of its travel until the circular outline of the leaves of the light source Iris are as sharp in focus (with the object) as possible. Open the light source iris all the way. This establishes condition II.

Aberrations:

Spherical aberration: Edges of a spherically curved lens will produce greater refraction than the center. This is corrected by combining lens elements such that one lens compensates for the spherical aberration of another.

Chromatic aberration: Refraction depends upon the wavelength of light. A single lens does not focus all colors at the same points. Combinations of lens elements of different refractive indexes are used. Achromatic objectives achieve adequate correction for ordinary purposes. High level resolution demands more correction, however, and apochromatic objectives made up of several elements of glass and the mineral fluorite are used.

Other aberrations have been nearly eliminated by manufacturers such as astigmatism, coma, curavature of the field and distortion; therefore, they are of little concern to the microscopist.

Filters

Blue and green filters increase the resolution by absorbing the longer light rays of the spectrum and transmitting only the shorter ones.

> = 435 to 490 blue
> 490 to 535 green

Filters can also increase the contrast or detail of a specimen.

Staining of Microsporocytes

1. Place anthers in a drop of staining fluid on a clean slide. After cutting them transversely, gentle pressure repeatedly applied with a curved needle will force the contents out into the drop.

2. Stir the drop vigorously with a rusty needle to separate the pollen mother cells and also to add iron to the stain. Note that the stage of meiosis is most easily recognized in cells at or near the edge of the drop. If not at the desired stage, the drop may be wiped off.

3. Remove the anther pieces by picking them up between the points of the curved needles.

4. Add a clean coverslip. The drop of stain should be no larger than sufficient to barely fill to the edge. For maize sporocytes no pressure is needed. For certain species, e.g. tomato, considerable pressure is needed to flatten and spread the cells. For these cases, squash the slide between blotting paper.

5. Heat the slide by passing it back and forth over an alcohol flame. Heat enhances the contrast between chromosomes and protoplasm. For maize, results are best when heated almost to boiling. Repeat the heating until no further improvement in contrast is noted. In maize, this degree of heating for fresh material or in fixative only a short time tends to remove the protoplasm.

6. The slide may be examined immediately under a microscope. Use light passed through a glass ground on both sides plus a green filter.

Taken from R.L. Phillips in <u>Staining Procedures</u>, ed. George Clark, 4th edition, Williams & Wilkins. 1981.

**Disjunction Frequencies In Interchange Heterozygotes
Using Microspore Quartet Analysis (revised 1983)**

One method of determining the frequencies of alternate, adjacent-1, and adjacent-2 modes of disjunction in an interchange heterozygote in corn (<u>Zea mays</u> L. 2n=2x=20) is based on the observation that the number of nucleoli per microspore depends on the number of nucleolus organizers in the microspore.

Normally, corn has a single nucleolus organizer region (NOR) on the short arm of chromosome 6. The absence of an NOR is reflected in a diffuse state of the nucleolar material in the microspore (i.e. no organized nucleolus), the presence of one NOR results in a single, normal nucleolus; and the presence of two NORs results in two nucleoli, except where they fuse to form one. The fusion of two nucleoli to form one is always discernible, however, since for every microspore with the potential of two NOR's there will be one microspore in the same quartet with dispersed nucleolar material. Thus, the absence of an organized nucleolus in one microspore of the quartet indicates that one of the remaining microspores should have two NORs.

Burnham (1950. Genetics 35:446-481) utilized observations to determine in maize the relationship between the frequencies of the three modes of disjunction and (1) and length of the interstitial (breakpoint-centromere) segments and (2) the pairing configuration. Interchanges involving chromosome 6 were utilized since chromosome 6 carries the single nucleolus organizer of maize.

The main purpose of this lab is to determine the frequencies of alternate, adjacent 1, and adjacent 2 modes of disjunction.

Pachytene Configuration of Stock   T5-6c

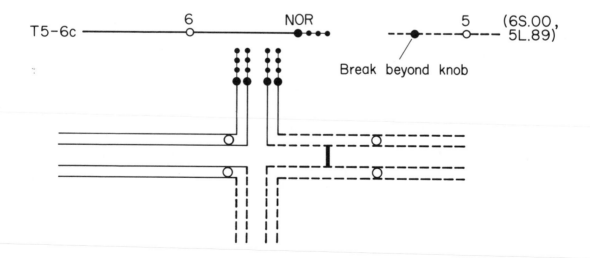

As mentioned in class, the meiotic configurations of interchange chromosomes vary considerably. What are some factors that affect these configurations?

Table 1 indicates that the relative frequencies of the different modes of disjunction depend upon the configuration of the interchange chromosomes and crossing over in the interstitial segment. For each unique interchange it is important to determine the relative frequencies of ring vs. chain formation. In corn this is easily accomplished at the diakinesis stage of meiosis.

Table 1

Configuration	C.O. in interstitial segment	Ratios of segregation types		
		alt	adj-1	adj-2
1. Ring	None	2	1	1
2. Ring	Yes	1	1	rare
3. Chain	None	1	1	rare
4.	Yes	1	1	rare

Assuming no crossing-over in the interstitial segment:

Modes of disjunction	Quartet Type(s)	Microspores
1) Alternate 1 & 2	no diffuse:	all 4 variable
2) Adjacent 1	2 diffuse:	all 4 abort
3) Adjacent 2	no diffuse:	all 4 abort

Assuming crossing-over in the interstitial segment:

Modes of disjunction	Quartet Type(s)	Microspores
4) Alternate 1 & 2	1 diffuse:	2 viable, 2 abort
5) Adjacent 1	1 diffuse:	2 viable, 2 abort
6) Adjacent 2	1 diffuse or 0 diffuse	all 4 abort / all 4 abort

Notes:

1) 2 nucleoli can fuse to form 1 nucleolus. This can be deceptive at first. The important observation to make is how many members of the quartet have diffuse nucleoli.

2) Adjacent II disjunction probably does not occur following a cross-over in the interstitial region.

3) Assuming no crossing-over in the interstitial segments, quartets resulting from Alternate and Adjacent 2 disjunction are indistinguishable even though the final products (microspores, pollen) have drastically different fates.

4) Assuming no crossing over in the interstitial region (and no 1 diffuse type quartet) the frequency of Adjacent I disjunction can be easily derived.

5) Assuming crossing over within an interstitial segment, quartets resulting from Alternate, Adjacent I and Adjacent II disjunctions are indistinguishable.

6) The frequency of 1 diffuse type quartets reflects the frequency of crossing over in the interstitial segments.

Estimating the Disjunction Frequencies

After observations have been made on the frequencies of the different quartet types and the actual percent of aborted pollen, it is possible to predict the expected percent pollen abortion and to estimate the frequencies of the modes of disjunction.

For example:

observed pollen abortion: 50.2%

Quartet Types

	0 diffuse	1 diffuse	2 diffuse	Total
counts	172	52	115	339

For the moment, ignore the diffuse quartet type. This class results from crossing over of the interstitial segment.

Frequency of Adj. I disjunction: 115 + 172 = 0.401 or 40.1%

This is a minimum estimate of Adj. I disjunction because it is likely that some of the 1 diffuse quartet types result from Adj. I disjunction. To predict pollen abortion, recall the following relationships:

        0 diffuse = 4 viable pollen grains
        2 diffuse = 4 aborted pollen grains
        1 diffuse = 2 aborted and 2 viable pollen grains

Predicted pollen abortion (assume no adjacent 2 disjunction):
$$\frac{(115 \times 4) + (52 \times 2)}{339 \times 4} = 0.416 \text{ or } 41.6\%$$

observed pollen abortion - predicted pollen abortion = excess pollen abortion

        50.2%   -   41.6   = 8.6%

This excess pollen abortion can be due to: 1) the genetic background of the stock (certain corn inbred lines have a characteristic level of percent pollen abortion unrelated to the presence of the heterozygous interchange), 2) the occurrence of Adjacent 2 disjunction which would result in 4 aborted spores even though each member of these Adjacent 2 quartets has 1 nucleolus (0 diffuse). These spores abort because they carry a chromosome that is both deficient and duplicate (draw these chromosomes).

Given the estimation of 8.6% excess pollen abortion, the following calculations can be performed to estimate the frequency of Adjacent 2 disjunction (assuming crossing over in the interstitial segments).

63

8.6% (or 29) of the total number of quartets are the result of Adj. 2 segregation and appear as "no diffuse" quartets.

The frequencies of segregations in the non-crossover quartets [1 diffuse (c.o. type) quartets are not included since they may arise via all three modes of disjunction] were:

Alternate	Adj.-1	Adj.-2	Total
143(172-29)	115	29	287
50%	40%	10%	

Calculate the frequencies of the three modes of disjunction for the stocks used in this lab.

Cytology of a Paracentric Inversion

OBJECTIVES

- To recognize meiotic abnormalities (bridges, fragments) in maize sporocytes

- To understand how cytological data can be used to explain observed genetic behavior.

MATERIALS

Maize sporocytes from a line heterozygous for a paracentric inversion, 4 in 7a.

DISCUSSION

A. Observable differences in meiosis

    1. Pachynema
    2. Diakinesis
    3. Anaphase I and II (handout)

B. Behavior of bridges

C. Behavior of fragments

D. Consequences of the meiotic behavior, cytologic vs. genetic results

E. Collecting and using cytological data

SOME QUESTIONS TO CONSIDER DURING THIS LAB:

How is the location of inversion breakpoints determined?

What kinds of chromosomes other than heterozygous paracentric inversions would cause bridges and fragments at meiosis?

How does the size of the loop affect the length of the fragment?

What effect does the position of the crossover within the loop have on the products formed?

What kind of sterility (pollen or ovule) would be expected in this line?

What does the frequency of sterility depend on?

Can the observed frequency of bridges be used to predict % ovule abortion?

LAB 5 - CYTOLOGY OF A PARACENTRIC INVERSION

How would you collect and report quantitative data for this?

What assumptions must be made in drawing conclusions from these data?

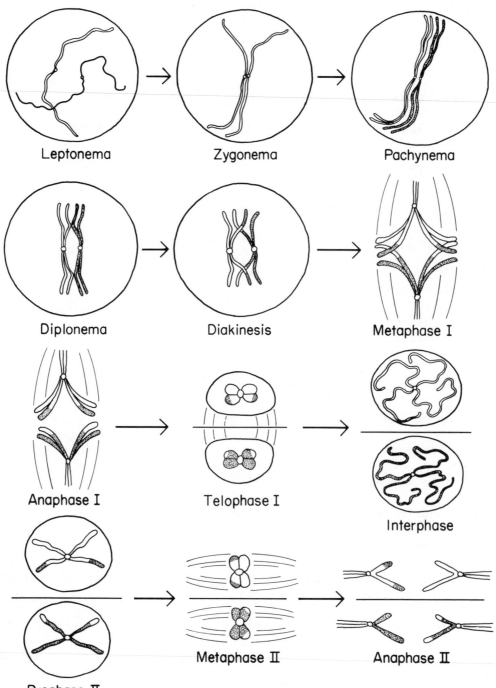

REFERENCE

Burnham, _Discussion in Cytogenetics_, Ch. 3.

ANAPHASE I and II configurations. Depending on the location and number of crossovers, bridges and fragments may be seen at anaphase.

A. If a crossover occurs <u>outside</u> the inversion loop.
   1. in the distal region— no bridges or fragments
   2. in the interstitial region— no bridges or fragments

B. If a crossover occurs <u>in</u> the inversion loop, several different patterns of bridges and fragments must be formed.

1. SCO—single crossover

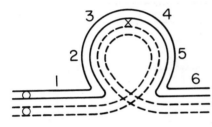 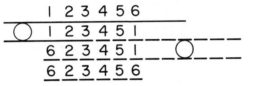

2. DCO—double crossover

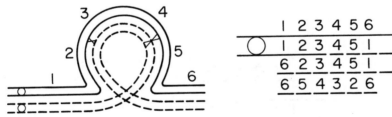

67

c. four-strand

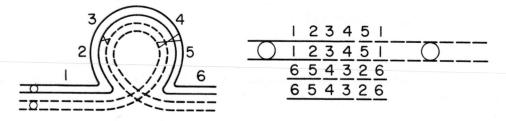

C. If there is a crossover in the <u>interstitial region</u> and one in the <u>inversion loop</u>, things become very interesting...

1. DCO, two-strand

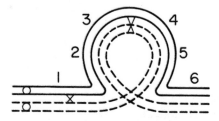

2. DCO, three-strand

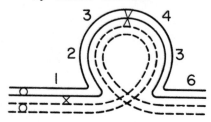

3. DCO, four-strand

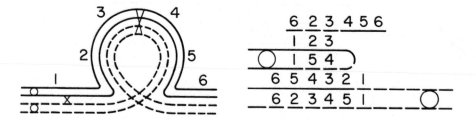

D. Consider other possibilities: what would happen if there were one crossover in the interstitial region and a DCO within the loop?

<u>Upper</u>. Majoric pethyrene configurations in an individual heterozygous for a paracentric inversion. The numbers indicate the positions at which various crossovers are assumed to occur, the results of which are listed in

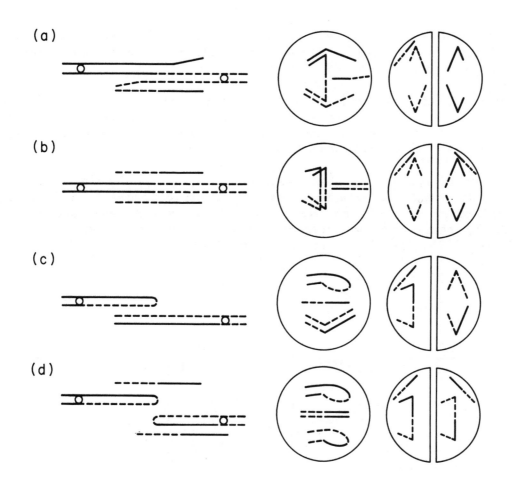

(a)

(b)

(c)

(d)

Lower. A, B, C, D are representive types of chromatids and anaphase configurations resulting from crossing over. The chromatids are shown on the left, followed on the right by the expected anaphase I and II configurations. The crossover chromatids are made up of parts of the two homologous, depending on the positions of the crossovers. The ones diagrammed are schematic and do not necessarily correspond to the stated crossovers. The ones diagrammed are schematic and do not necessarily correspond to the stated crossover points.

A. A single crossover within the inversion, e.g., at 3 or 4, results in a dicentric and an acentric chromatid and a bridge plus fragment at anaphase I.

B. A 4-strand double crossover within the inversion, e.g., at 3 and 4, results in two dicentric and two acentric chromatids and two bridges plus two fragments at anaphase I.

C. A crossover in the interstitial segment at 6, and certain ones within the inversion, e.g., a single at 4 or 3 results in a dicentric with sister centromeres and an acentric chromatid. At anaphase I there is a fragment but no bridge, but at anaphase II there is a bridge.

D. A crossover in the interstitial segment at 6, and a double at 3 and 4 produce a dicentric at each pole and two acentric fragments at anaphase I, and a bridge in each of the two cells of the dyad at anaphase II. At anaphase I, there are two fragments but not bridge. (Modified from McClintock, 1938, Fig. 3, p. 10, Mo. Res. Bull. 290).

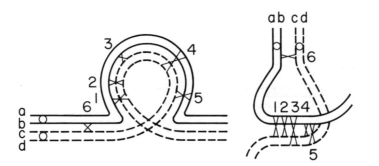

Kinds of chromatids produced by various crossovers in a paracentric inversion heterozygote. The configurations expected at anaphase I and II are included. All dicentrics have non-sister centromers unless indicated otherwise.

c.o. in inversion	Constituion of the strands after c.o.		anaphase div. 1	anaphase div. 2
	dup. & defic. chromatids	chromatids with full complement		
A. No crossover in any single	dicent. dup. & def.; accent def. & dup.	N In	bridge & frag.	no bridge
1,2=2 str. double	---	2N 2In	no bridge	"
1,3=3 "	dicent. dup. + def.; accent. def. & dup.	N In*	bridge & frag.	"
1,4=3 "	dicent. dup. & def.; accent def. & dup.	N* In	" "	"
1,5=4 "	2 dicent. dup. & def.; accent 2 def. & dup.	-- --	2 bridges & 2 frag.	
B. Crossover in 6 and single at				
1=2 str.	dicent. dup. & def.; accent. def. & dup.	N In	bridge & frag.	no bridge
3=3 "	dicent. dup. & def. (sister cent.); accent. def.& dup.	N In	frag., no bridge	bridge in 1 cell
4=3 "	dicent. dup & def. (sister cent.); accent def. & dup.	N In	" "	"

5=4 "	dicent. dup. & def.; accent.	N	In	bridge & frag.	no bridge
double at**					
1,2	--- ; ---	2N	2In	no bridge	no bridge
1,3	dicent. dup. & def.; accent. def. & dup.	N	In*	bridge & frag.	"
1,4	dicent. dup. & def.(sister cent.); accent. def. & dup.	N*	In	frag., no bridge	bridge in 1 cell
1,5	2 dicent. dup. & def.; 2 accent. def. & dup.	--	--	2bridges & 2 frag.	no bridge
3,4	2 dicent. dup & def. (sister cent.); 2 accent. def. & dup.	--	--	2 frag., no bridge	bridge in both cells

*Those chromatids have a double crossover between the inversion breaks. Two of the chromatids from 2-strand doubles also have a double crossover.

** Only part of the possible combination are shown.

Squash Technique For Maize Root Tips And Suspension Cultures

A. *Zea* mays

1. Germinate seeds -- collect root-tips at about 2-3 days.

2. Pre-treat in 0.3% colchicine for 3-4 hr to arrest mitotic division and shorten the chromosomes.

3. Kill and fix in Farmers solution (3:1, 95% ETOH: glacial acetic acid) for 12-24 hr.

4. Transfer root tips to 70% ETOH and store at 2-3°C until needed.

5. Soften root-tip by procedure a or b.

   a. hydrolyze in 1 N HCl at 60°C for 5-10 min.

   b. 1% pectinase and 1% cellulase for 45 min to 1 hr.

6. Apply a small drop of propiono-carmine stain to a slide and macerate the root-tip using iron needles.

7. Apply a coverslip and squash as demonstrated.

8. Heat gently and observe under low power.

Cytology of cell suspension culture

1. Treat 3-day-old suspension culture with 0.2% colchicine for 8-10 hr.

2. Fix in 3 parts 95% ethanol:1 part glacial acetic acid overngiht at room temperature.

3. Wash the fixed cells twice with 0.1 M sodium acetate (pH 4.5).

4. Digest cells with 2 volumes of 0.5% (W/V) cellulae and 0.5% pectinase.

5. Rinse cells twice with 45% acetic acid and store in 45% acetic acid at 4°C until needed.

6. Apply a small drop of cells to a slide and stain with a drop of propionic carmine for microscopic observation.

IN SITU HYBRIDIZATION WITH MAIZE MEIOTIC CELLS

R.L. Phillips and A.S. Wang
Department of Agronomy and Plant Genetics
University of Minnesota
St. Paul, MN 55108

Reprinted from Maize for Biological
Research W.F. Sheridan, ed.
Copyright 1982, Plant Molecular Biology
Association, P.O. Box 5126,
Charlottesville, VA 22905

In situ hybridization of RNA to maize chromosomes allows the localization of specific nucleotide sequences. The procedure given here will detect the location of genes repeated 50 times or more at a particular site. We have employed this method to further localize the 17-26 S ribosomal RNA genes within the nucleolus organizer region of chromosome 6 (Phillips et al., 1979) and to confirm the location of the 5S rRNA genes to the long arm of chromosome 2 (Mascia et al., 1981).

Slide Preparation:

1. Fix freshly collected microsporocytes in 3 parts 95% ethanol and 1 part glacial acetic acid for 1-3 days at room temperature.

2. Use one of three anthers in a floret to identify appropriate meiotic stage and keep remaining anthers in a vial containing fixative.

3. Squash 3-5 anthers on an acid-cleaned side in a drop of 45% acetic acid and add an acid-cleaned cover slip.

4. Place slides on dry ice for 5-10 min. Use a razor blade to flip off the cover slip.

5. Keep slides in 2 x SSC (1 x SSC: 0.15 M NaCl and 0.015 M Na citrate, pH 7.4).

Denaturation of DNA:

1. Dissolve RNase in 2 x SSC (0.2 mg/ml) and heat at 80°C for 10 min.

2. Digest the meiotic cells with the heat treated RNase for 2 hr at room temperature (or 1 hr at 37°C) to remove endogenous RNA which may compete with the hybridizing RNA.

3. Remove the RNase by washing 3 times with 2 x SSC.

4. Treat the slides with 0.2 N HCl for 20 min, at room temperature.

5. Wash the denatured slides 3 times with 2 x SSC.

RNA-DNA Hybridization:

1. Place 20 1 of ^{125}I-RNA or ^{3}HcRNA on each slide (conc.: 0.25 g/ml with 10 g E. coli RNA as carrier). RNA is dissolved in 4 x SSC and mixed with an equal volume of formamide resulting in a final solution that is 2 x SSC and 50% formamide

2. Cover the slides with an acid-cleaned cover glass and place in a pan containing 2 x SSC or deionized water and support the slides with glass rods.

3. Seal the pan with aluminum foil and keep in 45°C oven overnight (16 hr.).

4. Float the cover glasses off the slides with 2 x SSC and wash three times with 2 x SSC.

5. Digest with 80°C heat treated RNase (0.2 mg/ml) for 2 hr at room temperature or at 37°C for 1 hr.

6. Wash the slides 3 times with 2 x SSC and keep in 2 x SSC.

Autoradiography:

1. Melt Kodak NTB-2 emulsion (112 ml) at 45°C and dilute with two volumes of distilled water in absolute darkness.

2. Pour 30 ml of the diluted emulsion in a small container and save the remainder for future use.

3. Dip the slides into the emulsion and withdraw slowly. Place slide vertically in a rack for 10 min and wipe the wet end.

4. Place the emulsion-coated slides in a light-tight plastic slide box with Dririte inside. Seal the box with black electrical tape, wrap with aluminum foil, and store at 4°C to expose.

5. After proper exposure time (2-4 days), develop the slides in Kodak D-19 developer for approximately 15 sec and rinse 3 times with distilled water.

6. Fix slides in Kodak fixer for 1 min and wash with distilled water for 20 min through several changes of water.

7. Stain the slides with 5% Giemsa stain (Harleco) before drying (dilute the Giemsa solution with 0.01 M phosphate buffer at pH 6.8).

References

Mascia, P.N., I. Rubenstein, R.L. Phillips, A.S. Wang, and Lu Zhen Xiang. 1981. Localization of the 5S rRNA genes and evidence for diversity in the 5S rDNA region of maize. Gene 15:7-20.

Phillips, R.L., A.S. Wang, I. Rubenstein, and W.D. Park. 1979. Hybridization of ribosomal RNA to maize chromosomes. Maydica 24:7-21.

MAIZE PROTOPLAST/LIPOSOME CAT ASSAY

Assay of Chloramphenicol Acetyl Transferase

Materials

Small test tubes
15 ml Covex tubes
15 ml Falcon plastic EFG tubes
Tris
Acetyl-CoA (Sigma or PL Biochemicals)
Chloramphenicol Acetyl Transferase (PL Biochemicals)
1-^{14}C-Chloramphenicol 45 mCi/mmole # CFA.515 (Amersham)
PMSF
Leupeptine (Sigma or Boehringer)
X-ray film XAR-5 (Kodak)
Chloroform
Methanol
EN^{3}HANCE (New England Nuclear)
Ethyl Acetate
Speed-Vac
Pasteur pipets
37°C H$_2$0 bath
13179 Silica gel wo indicator plates (Eastman)
  or Baker-fex Silica gel 1B (J.T. Baker Chem)
TLC tank lined with WHATMAM 3 MM, containing Chloroform 95%, methanol 5%
Chromist spray units (for EN^{3}HANCE) - Gelman 51901
  Curtin Matheson Scientific Cat. # 343-848
  reservoirs -Gelman 78068. CMS Cat. # 352-848
Sonicator
60°C H$_2$0 bath

Stock Solutions

Leupeptine 25 mM (11.85 mg/ml H$_2$0) - stable for 1 week
PMSF (100 mM in MeOH)
TE 250/10 - 250 mM Tris 7.5
            10 mM EDTA
Acetyl-CoA   10 mM (8.09 mg/ml H$_2$0) - stable at 4°C

Note: We orginally attempted the assay of Van Montegu and Schell, however, we were unable to extract the chloramphenicol and derivatives with ethyl acetate.

Method

1. Collect cells in 15 ml plastic cfg tube by slow cfg, remove
    supernatant.

2. Resuspend cells in 1.0 ml TE 250/10
                    10 ul PMSF

20 ul leupeptine

3. Mix and sonicate a few seconds to disrupt cells (microtip – #1 x 3"; bath sonicator – "30").

4. Heat to 60°C x 10', cfg 10' at "10K x g.

5. Transfer super to 15 ml Corex. Add 2.5 ul (0.5 ul (0.5 uCi) ^{14}C-CAM, 50 ul acetyl-coa. vortex.

6. Incubate 37°C x 30' – NOT LONGER.

7. Extract w/equal volume ethyl acetate. Separate layer by cfg (5' @ 7K x g).

8. Remove organic (upper) layer. (This can be frozen)

9. Evaporate to dryness, resuspend in 10 ul Ethyl Acetate.

10. Spot onto TLC and develop in 95:5 HCCl$_3$:MeOH.

11. Spray w/EN^{3}HANCE, expose onto X-ray film "1 week.

12. Develop film.

Control includes enzyme (PL-biochemicals) plus substrates.

References

Herrea-Estrella, et al. Nature <u>303</u>:209-213 (1983) and references therein.

DiNocera and Dawid PNAS <u>80</u>:7095-7098 (1983).

MAIZE MITOCHONDRIAL DNA PURIFICATION (modified Kemble procedure)

1. Grind 5-10g lots in mortar for 30 sec in 3 vol/wt Buffer A.

2. Filter through 4 layers cheesecloth and 1 layer miracloth (don't squeeze miracloth).

3. Centrifuge Sorvall SS-34, 50-ml tubes, 10 min @ 3,000 rpm.

4. Centrifuge supernatant 10 min @ 10,000 rpm.

5. Resuspend pellet in 0.5 ml/gm Buffer G using bristle brush.

6. Centrifuge 10 min @ 3,000 rpm.

7. Supernatant plus 0.1ml/10 ml 1M $MgCl_2$ + 5 ul/gm of 2 mg/ml DNase. Mix and incubate 60 min @ room temp.

8. After digestion layer 20 ml Shelf Buffer <u>under</u> digestion mix.

9. Centrifuge 20 min @ 9,000 rpm.

10. Resuspend pellet in 10 ml Shelf Buffer, centrifuge 10 min @ 10,000 rpm. Repeat step 11.

11. Resuspend in 0.1 ml/gm Lysis Buffer + 25 ul/gm of 10% SDS + 1.5 ul/gm of 20 mg/ml Proteinase K. Stopper, mix and incubate 60 min @ 37ºC with gentle shaking.

12. After incubation transfer to 15 ml siliklad Corex glass tube, add 0.3 ml 8M NH_4OAc + 3 ml Phenol + 3 ml Chloroform-octanol (24:1), stopper with cork and mix by hand.

13. Centrifuge 5 min @ 7,000 rpm.

14. Transfer upper phase (aqueous) to clean Corex tube, add Phenol and chloroform octanol, centrifuge as in step 16. Repeat Step.

15. Transfer upper phase to 50 ml centrifuge tube, measure vol, add 0.4X that volume in 7.5u NH_4 acetate and allow to stand on ice for 5 min. Add 2.0X total volume (w/NH_4Ac) 100% EtOH, mix gently, stopper and store overnight in -20ºC freezer.

16. Centrifuge 10 min @ 8,000, mark the pellet side of tube.

17. Decant and add 10 ml 70% EtOH to pellet, sit in ice for 10 min and centrifuge as in step 20. Repeat.

18. Air dry final pellet at room temp.

19. Take pellet up in 5 ul/gm TE buffer on ice for 10 min. Transfer to 1.5 ml Eppendorf tube and repeat step.

20. Combine, label tube and freeze @ -20ºC.

Buffers for Mitochondrial DNA Isolation

Buffer A	mol. wt.	g/2l
.5 M Sucrose	342.2	342.3
.05 M Tris (Trizma base)	121.1	12.11
.005 M Na$_2$ EDTA	372.2 (+2H$_2$O)	3.722
.1% BSA		2.0

pH to 7.5 w/HCl, bring to volume and filter through Whatman #1

Buffer G	mol. wt.	g/l
.3 M Sucrose	342.2	102.7
.05 M Tris (Trizma base)	121.1	6.055

pH to 7.5 w/HCl, bring to volume and filter

Shelf buffer	mol. wt.	g/l
.6 M Sucrose	342.3	205.4
.01 M Tris (Trizma base)	121.1	1.211
.02 M Na$_2$ EDTA	372.2	7.444

pH to 7.2 w/NaOH, bring to volume and filter

Lysis buffer	mol. wt.	g/200 ml
.5 M Tris (Trizma base)	121.1	1.211
.01 M Na$_2$ EDTA	372.2	.744

pH to 8.0 w/HCl, bring to volume

TE buffer	mol. wt.	g/l
.01 M Tris (Trizma base)	121.1	1.211
.001 M Na$_2$ EDTA	372.2	.372

pH to 7.5 w/HCl, bring to volume

MAIZE GENOMIC DNA

NOTE: We give 3 different methods below for the isolation of long
nuclear DNA molecules suitable for production of genomic libraries
in lambda. In class we will use method A.

Method A

Modified by Mike Zarowitz from a modification by Steve Wietgrefe from
a modification by Tom Guilfoyle from Hamilton, Kunsch, and
Tempelli (Analytical Biochemistry, 49, 48-57, 1972). The DNA
isolation protocol is from Kislev and Rubenstein (Plant
Physiology, 66, 1140-1143, 1980).

Materials

50 - 100 gm Black Mexican Sweet (BM7) cells (best if fresh)
Tris
MgC12
EDTA
NaCl
3M Sodium acetate
Sucrose
B-mercaptoethanol
Ethidium Bromide
Percoll
Sodium dodecyl sulfate (SDS)
Proteinase K
RNase A
Phenol (equilibrated with STE, below)
Isopropanol
Ethanol (95% and 70%)
Glass Rods for spooling DNA
10 ml pipets
Platform rocker or slow rotary shaker
37°C incubator
-15°C freezer
Chloroform:isoamyl alcohol 24:1
Buchner funnel, fast flow filter paper, vacuum apparatus
cheesecloth
miracloth
diethylether
Azure C (Sigma)
Sodium Azide
Triton X-100
A swinging bucket rotor for a Sorvall RC2B type fuge
Polytron
30 ml Corex tubes plus adaptors for centrifuge
"Magic Marker"-type felt pens
Microscope plus accessories

Note: For late log – early stationary BM7 suspension cultures, one
      can expect 0.35 to 0.5 gm of cells per milliliter of culture.
      Doubling time of BM7 is about 60 hr.

Stock Solutions

Ether (cold)

AB: 10 mM Tris pH 7.6
     1.14 M Sucrose
     5 mM MgCl2
     10 mM B-ME
     10 ug/ml Ethidium Bromide (omit for functional nuclei)

100% Percoll:
     20 ml Percoll
     0.2 ml 1 M Tris pH 7.2
     0.1 ml 1 M MgCl2
     16 ul B-ME
     6.95 gm Sucrose (to make 1M)

Azure C stain:
     0.1% Azure C in 250 mM sucrose, 0.02% NaN3
     (can also use 0.5-1% acetocarmine in 45% acetic acid)

Proteinase Digestion Medium (PDM):
     10 mM Tris pH 7.5
     10 mM EDTA
     10 mM NaCl
     0.5% SDS

STE:
     1 mM Tris pH 8.6
     100 mM NaCl
     1 mM EDTA

TE/10:
     1 mM Tris pH 7.6
     0.1 mM EDTA

Proteinase K:
     1 mg/ml in PDM

RNAase:
     4 mg/ml; boiled 5 min to destroy DNases.

Procedure

All procedures are performed at 0-4°C.

1. Collect c. 50 gm of fresh BM7 cells by filtration, transfer to a beaker. Add 200 ml diethylether, swirl or stir 2-3', pour off ether and blot excess.

   Note: Alternatively, the appropriate plant tissue can be chilled and then cut into small pieces with a razor blade, or frozen in liquid nitrogen and then ground in a Waring Blendor (high setting for 30-60").

2. Add 2-4 volumes (c. 100 ml) of fresh AB. Grind in Polytron at setting #5 for 30-45 sec. In our hands, this produced nuclei and few intact cells. This can be checked by observing an Azure C stained aliquot with a microscope.

3. Filter through 4 layers of cheesecloth, then one layer of Miracloth.

4. Divide filtrate into 30 ml Corex tubes. Cfg 12' in swinging bucket rotor at 500-1000 x g.

5. **DO NOT DISCARD SUPERNATANTS** Resuspend pellets in a TOTAL volume of 20-25 ml AB. Check for nuclei with Azure C stain or with 0.5-1% acetocarmine in 45% acetic acid. If nuclei have not pelleted, re-spin supernatants.

6. Resuspend nuclei in AB plus 0.25% Triton X-100. Cfg 12' at 1000 x g.

7. Resuspend pellet in 10-20 ml AB (in a 30 ml Corex tube). 0.25 Triton X-100 is optional at this step.

8. Set up step gradients, 5 ml each of 30-50-70-90% Percoll (mark layers). % Percoll is prepared by diluting the 100% Percoll with either AB or with a solution identical to the 100% Percoll stock except 20 ml water was used instead of the 20 ml Percoll. To fine tune this step, use a step that will just retard the nuclei. That way the nuclei will band and the starch will pellet.

9. Cfg, 30' at 7,000 x g in a swinging bucket rotor.

10. Identify nuclei (with Azure C stain) and pool. Determine nuclei concentration.

    Dilute 5X with AB.

11. Pellet by cfg at 650-1000 x g for 10 min. Can repeat Percoll gradient if necessary.

12. Wash 2X with 30 ml AB to remove residual Percoll. 0.25% Triton X-100 is optional in either wash.

Long Term Storage:

Resuspend nuclei pellet in 1-3 ml of 250 mM sucrose, 20 mM HEPES pH 7.8, 5 mM MgCl2, 1 mM DTT, 50% (v/v) glycerol; at -79 degrees C. RNA polymerase activity is stable for many months (no detectable loss) even after multiple cycles of freezing and thawing.

DNA Isolation from Nuclei

1. Resuspend nuclei in 8 ml of PDM. The DNA is then mildly sheared (will still be larger than 100 kb) by GENTLY pipetting the solution up and down 3 times in a 10 ml pipet.

2. Add 4 ml of Proteinase K stock (1 mg/ml) and gently mix at 37°C overnight.

3. Cfg at 20,000 x g for 20 min at 4°C.

4. Extract the supernatant three times with a 1:1 mixture of phenol/STE : chloroform/isoamyl alcohol. Extract once with cholorform/isoamyl alcohol.

5. Add two volumes of 95% ethanol, one-tenth volume of 3 M sodium acetate, and precipitate the DNA overnight at -15°C. Collect the DNA by cfg at 12,000 x g for 20 min.

6. Resuspend pellet in 2 to 5 ml of TE/10. Add RNAase to a final concentration of 40 ug/ml and incubate on ice for 30 min.

7. Extract as in Step 4.

8. Add 0.1 volume of 3 M sodium acetate and overlay with 0.54 volume of isopropanol. The DNA is then collected by spooling onto a glass rod, rinsed with 70% ethanol, and resuspended (overnight) at 4°C in 1 to 3 ml of TE/10.

Method B   Shure, Wessler, & Fedoroff 1983 Cell 35:225-233

For 20 gm of tissue (highest yields from seedlings or the young, inner leaves of older plants)

> Lysis Buffer - for 400 ml - ice cold
> 168 gm urea
> 28 ml 5 M NaCl
> 20 ml 1 M Tris-Cl pH 8.0
> 16 ml 0.5 M EDTA
> 20 ml 20% Sarkosyl
> 20 ml phenol

1. Powder tissue in liq. N_2 - mortar and pestle.

2. Add 20 gm powder to 85 ml lysis buffer in flask - swirl.

3. With a glass rod break up clumps.

4. Add an equal volume of phenol:chloroform:isoamyl [25:24:1, equilibrated with TE (10 mM Tris pH 8, 1 mM EDTA)]. Add SDS to 0.5%

5. Swirl on a rotating platform 10-15 min room temp.

6. Spin in a clinical centrifuge at room temperature, 15 min at 3500 rpm.

7. Repeat extraction twice of aqueous phase with phenol:chloroform:isoamyl, with no SDS added.

8. To final aqueous phase - add 1/20 volume 3 M potassium acetate + 2 vol. ice cold ethanol - swirl - spool out chromosomal DNA with a glass rod - shake off excess ethanol. Dissolve in 14 ml 10 mM Tris (pH 8) 10 mM EDTA. (leave at 37°C about 1 hr then gradually increase temp to 67°C or until it all dissolves).

9. Add 0.95 gm/ml CsCl (13.3 gm/14 ml) + 300 ul 10 mg/ml EtBr. Spin in ultracentrifuge overnight in a vertical rotor, or 2 days in a fixed angle rotor.

10. Remove band either by dripping from the bottom, a needle in the side, or by suction from the top (your favorite method). Add 2 volumes of water, and extract the ethidium bromide three times with water saturated sec-butanol.

11. Dialyze overnight against TE. Concentrate if necessary with with sec-butanol. Remove the butanol with ether, and then boil off the excess ether by incubation at 37°C.

Yields - from 20 gm tissue (seedlings) about 1 mg DNA.

Method C

> NOTE: The molecular weight of the resulting DNA depends strongly on the health of the initial plant tissue. For cell cultures choose rapidly growing exponential phase liquid suspension cultures rather than stationary phase or callus cultures. For whole plants, choose young, fresh, material such as unopened leaves. From optimum tissue it is possible to obtain DNA in excess of 100 kb ds length.

HB
300 mM sucrose
5 mM $MgCl_2$
50 mM Tris pH 7.8
filter sterile

LYSIS BUFFER
50 mM Tris pH 8.0
50 mM EDTA pH 8.0
50 mM NaCl
2% Sarkosyl
400 ug/ml ethidium bromide

HBT
98 mls HB
2 mls Triton X100
filter sterile

MIRACLOTH FUNNEL
four layers of miracloth
in a funnel
autoclave

LIQUID NITROGEN

Preheat the lysis buffer to 65°C.

1. Collect liquid suspension cultures on filter paper in Buchner funnel under vacuum, weigh. For other tissues simply weigh. Drop into Dewar flask with liquid nitrogen. When frozen, pour off almost all the liquid nitrogen, then transfer tissue to Waring Blendor. [Do not freeze the tissue in the blendor as it will freeze the motor also]. Grind to a powder.

2. Add 3 ml of HB per gram of tissue. Grind 30 sec on high. Pour homogenate through miracloth funnel into centrifuge tube. Wearing gloves, gently gather up the miracloth and squeeze out liquid into centrifuge tube. Spin 2500 rpm for 15 min in refrigerated centrifuge (1000g).

3. Gently resuspend pellet in 70 ml of HB, respin at 2500 rpm for 15 min. Gently resuspend pellet in 35 ml of HBT, respin at 2500 rpm for 15 min. Gently resuspend pellet in 8 ml of HBT, respin at 2500 rpm for 15 min.

4. Gently resuspend pellet in 5 ml of lysis buffer, cover tube and gently invert to disperse pellet completely. (You can also disperse pellet by a very slow, careful, pipetting up and down in a plastic pipet). Incubate at 65°C for 30 min.

5. Add 0.95 gm CsCl per ml of lysate. Gently mix to dissolve. Spin 10,000 rpm for 20 min, save supernatant.

6. Load supernatant in ultracentrifuge. In order to balance the tubes, or dilute the extract, it is useful to prepare lysis buffer plus the correct amount of ethidium bromide. Spin 55,000 rpm overnight in vertical rotor. Collect band either by needle or tube puncture. The DNA should be viscous. In some plant preps there will be oils and starch at the top of the

gradient that bind EtBr; if there are lots of these they may even pull the EtBr out of the DNA, making it look like there is less DNA than there really is. It is frequently useful to combine bands from several tubes and respin, as a further purification.

7. Measure volume and then dilute the DNA bands with at least 2 volumes of water. Remove the ethidium bromide with 3 to 4 sec-butanol extractions. Recheck volume, making sure it is at least 3 times the original, and then precipitate with an equal volume of isopropanol, at -20°C for at least 1 hr. Spin down DNA, wash with 80% ethanol, spin down again. Dry pellet. Resuspend in TE by letting it sit at room temperature overnight.

NITRATE REDUCTASE ASSAY

Filner, P. 1966 Biochim. Biophys Acta 118:299-310
Muller, A., Grafe, R 1978 MGG 161:67-76

Nitrate reductase is an important enzyme in plant somatic cell genetics because it has been characterized biochemically, and because a positive selection exists for deficient mutants. The enzyme converts nitrate to nitrite, and also will convert chlorate to the toxic chlorite. Chlorate resistant lines therefore will be deficient in the enzyme (and require an amino acid supplement for growth).

Grinding Buffer

100 mM Hepes pH 7.5
10 mM DTT
chill on ice

100 mM KPO4 pH 7.5

Spectrophotometer at 540 nm

100 mM KNO_3

NADH
2 mg/2.7 ml

SULF:
1% Sulfanilamide in 3 N HCl

NAPHTH
0.02% Naphthylenediamine HCl

NITRITE:
KNO_2 - 0, 10, 20 ... 100 nanomoles/0.3 ml.

1. Collect early log phase cells on filter paper, weigh, add 2 ml grinding buffer per gm cells. Homogenize with mortar and pestle, or with glass/glass hand homogenizer. Spin 10,000 rpm, save supernatant.

2. Assay mixture:
   0.5 ml   KPO4
   0.1 ml KNO_3
   0.1 ml   NADH
   0.3 ml   Enzyme Supernatant (or 0.2 H_2O + 0.1 enzyme, etc.)
   Incubate 30 min at room temperature [the actual optimum for the enzyme is higher]

3. Add 1 ml Sulf and 1 ml Napth. Vortex, Let sit 15 min, read absorbance at 540 nm.

4. Controls: Assay without enzyme, without NO_3, without NADH.
   Standard: Assay known amounts KNO_2 in Mix without enzyme.
   1 unit = 1 nanomole/hr.   (specific activity = per mg protein)

This experiment will be coordinated with the liquid suspension culture stocks you are maintaining. The Class will be divided into 3 groups to test the effects of different media on the activity of nitrate reductase: (1) normal MS media, (2) no ammonia, only nitrate, (3) no nitrate, only glutamine.

PROTEIN EXTRACTION FROM CELL CULTURES FOR SDS GELS

Grinding Buffer
100 mM Tris pH 7.4
5 mM DTT
(0.1% SDS optional)

Acetone
at −20°C

2X and 1X SDS Sample Buffer

NOTE: The first protocol given below is a standard SDS boiling extraction to solubilize as many proteins as possible. This protocol usually requires an acetone precipitation step in order to concentrate the protein; a trichloroacetic acid precipitation does not work as well.
    The second protocol is a simpler protocol that works well for proteins that are soluble without SDS.

First Method — SDS extraction

1. Collect cell cultures by vacuum filtration. Weigh material. For whole plant material, slice into small pieces and weigh.

2. Add 3 ml grinding buffer (containing 0.1 % SDS) per gram of material, and place the tube containing the cell slurry in boiling water for 5 min. Homogenize with Polytron, blender, or other homogenizer. Spin 10,000 rpm for 20 min.

3. Save supernatant, measure volume. Add 9 volumes of chilled acetone per ml of supernatant. Mix. Sometimes a clot of gooey material that floats will form, which should be removed with forceps at this stage. Incubate at −20°C for 1 hr.

4. Spin 10,000 rpm for 20 min, save pellet. Resuspend in 1X-SDS Sample Buffer, usually in an volume in ml equal to 1/2 the original weight of material in gm. Before boiling the sample, spin out any insoluble material by a 15 min spin in a microfuge. Place tube in boiling water for 3 min, and let the tube return to room temperature. Load several different volumes in wells on a standard SDS gel.

Second Method — Soluble Proteins

1. Collect cell cultures by vacuum filtration. Weigh material. For whole plant material, slice into small pieces and weigh.

2. Add 1 ml grinding buffer (without SDS) per gram of material. Homogenize with mortar and pestle, or glass/glass homogenizer. Place in ultracentrifuge tubes and spin at 100,000g for 15 min.

3. Save the supernatant, and freeze for storage. Add an equal volume of 2X-SDS Sample Buffer, and place tube in boiling water for 3 min. Let the tube return to room temperature, and load several different volumes in wells on a standard SDS gel.

ANALYSIS OF PHOTOSYNTHESIS MUTANTS

Introduction

The photosynthetic electron-transport chain can be conveniently divided into three regions; the PS I and PS II complexes, organized around their respective reaction centers, and the so-called cytochrome f-b$_6$ complex. Additionally, the thylakoids contain two complexes not directly involved in electron transport; the light-harvesting chlorophyll a/b-protein complexes and the CF$_1$CF$_0$-ATP synthetase. Information concerning the modes of action, composition and regulated synthesis of these supramolecular entities has been obtained through the use of a variety of techniques. The use of genetic mutants altered in specific portions of the photosynthetic system can be particularly rewarding, since this approach utilizes the organism's own regulatory systems to define a physiological rather than an operational organization in the membrane.

Classically, mutational studies have employed the microalgae which can be manipulated with standard microbiological techniques. Recently, however, a wide range of mutants have been selected in maize and barley allowing for the extension of this method to higher plants. In this study we have utilized a mutant of <u>Zea</u> <u>mays</u> to investigate the composition of PS II and PS I. Those mutants are originally selected due to its increased fluorescence relative to normal plants with unimpaired photosynthetic electron transport.

Procedures

<u>Plant Material for Maize Photosynthesis Mutant Screening</u>

There are three different mutants to be planted which I have labelled A,B,C. Each package contains about 250 seeds treated with a fungicide. If there are to be eight different preparations of chloroplasts there should be almost 200 plants per preparation. This should yield 40 to 50 mutant plants which will be twice as many as needed.

Plant 250 seeds of each mutant 14 days before the time of the first laboratory and 250 more seeds of each mutant 9 days before the first laboratory. These should be planted in greenhouse flats in a potting soil mixture with about one inch spacing. The space is not critical but should leave enough room to easily screen the plants. Plants should be grown in a warm greenhouse (28 to 32°C). Treat plants with a high nitrogen fertilizer twice during this time or add a slow release fertilizer to the soil. Watch for red stems or brown leaf tips. This indicated lack of nitrogen. If the volume of soil is shallow, watch for afternoon drying when the plant leaves expand. On hot days they may need watering twice a day. If the plants seem to be growing too fast (ideal age would be to have three leaves for the first day of laboratory) you can put them in lower temperature or lower light intensity.

<u>Screening</u>

When the seedlings have grown uniformly to the 2- to 3- leaf stage (9 to 14 days) they can be screened in order to locate the one-quarter photosynthesis mutants. The material screened should segregate homozygous recessive photosynthesis mutants. The heterozygous can not be separated from the normal green plants. Most of the mutants which completely block photosynthesis can be identified at an older age since the lack of photosynthesis kills the seedlings. For photosynthetic studies we would like to select the mutants before they cause lethality of the seedling.

Some, but not all, mutant plants have an altered leaf pigmentation. Seedlings will appear slightly yellow-green, pale green or yellow. Many mutants have normal pigmentation but can be selected by changes in their level of chlorophyll fluorescence.

Fluorescent Screening

Seedling plants are examined in a dark room with a longwave UV lamp (peak output 366 nm). Protection for eyes is provided with UV safety goggles or clear plastic goggles. Illuminate the plants at a distance of 10-20 cm. Place the red plastic filter bewteen the plants and your eyes. Mark those seedlings which appear more red. The whole plant should be more fluorescent if it is a mutant, not just the leaf base. Later, rescreen the plant to see if the same seedlings were selected a second time. Record the number of plants selected as high fluorescent, the number with changes in pigmentation and the number of normal green plants.

Note if any plants are beginning to die. This occurs first in the upper half of leaf #2 and #3, then proceeds down the leaf.

Characterization of Mutants

Measurement of in vivo chlorophyll fluorescence can be used to see what portion of the light reaction is limited by the mutation. Cut off a 3-cm piece of leaf #2 (keep it damp on a paper towel) and place the leaf in the dark.

A filter fluorometer is set up which is a modification of the described in Meth. in Enzymol. 69:3, 1980.

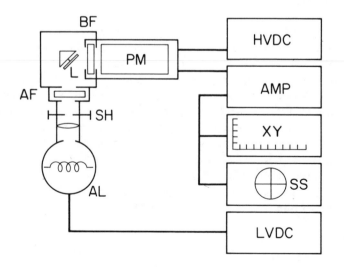

90

Test the baseline of the reading of the fluorometer without a leaf by turning on the fluorometer, turning on the lamp, turning on the recorder and opening the shutter. Record the response over 15-30 sec. Now place a normal green leaf in the fluorometer which has been dark adapted at least 3 min. Measure the kinetics of the fluorescence emission from the leaf. Measure the fluorescence of a mutant leaf from the same family. Repeat as many times as time and plant material permits.

The light-induced changes of fluorescence can now be examined to determine the function of photosystem I, II, or both in mutant leaves.

Fluorescence of isolated chloroplasts can be done using isolations from normal green or mutant leaves. The method is given later for electron transport measurements. Using chloroplasts, various electron donors and electron acceptors can be provided to further characterize the portion of the electron transport chain altered by mutation. This will be optional if time is available.

Electron Transport

Isolation of chloroplasts -- Leaves are harvested in the morning, if possible, to reduce starch content and weighed. Four to six grams of leaves are washed twice and stored cold until ready for use. This procedure yields broken chloroplasts and thylakoid membranes.

Leaves are ground thoroughly in a cold (in ice) mortar and pestle. Thirty to 60 ml of isolation buffer (pH 8.0 checked) is used for the grinding (see Appendix for make-up of reagents). The ground leaves are transferred to a layer of Miracloth in a funnel and the filtrate collected in cold centrifuge tubes.

Centrifuge (cold) the preparation 2,000 xg for 10 min to sediment the chloroplasts. Discard the supernatant. If time permits and the chloroplast pellet is large enough it can be resuspended in 25 ml of the resuspending medium, filtered with Miracloth and centrifuged again. Otherwise, resuspend in a minimal volume of resuspended medium (pH 8.0). This should be done cold with a glass rod and usually requires 2 ml. Store on ice. Chloroplasts are useable for 2 to 3 hr. Storage time can be extended by adding 0.1% BSA to the resuspending medium.

Chlorophyll Determination

Stir the chloroplast suspension, take a 50 ul sample and dissolve in 5.0 ml of 80% acetone. When the Chl is completely dissolved, fill a centrifuge tube, stopper, and centrifuge the sample in a clinical centrifuge 3 to 5 min. Decant samples into cuvettes to measure chlorophyll.

Zero spectrophotometer, and measure absorbance at 645 and 663 nm. If a scanning instrument is available, a spectrum of mutants and normal green plants could be recorded. Use the following from Arnon (Arnon, D.I. 1949 Plant Physiol. 24, 1-15) to determine: Total Chl; Chl a; Chl b and a/b. Values are in mg Chl per ml of chloroplast suspension.

$$(127 \times A_{663}) - (2.69 \times A_{645}) = Chl\ a$$

$$(22.0 \times A_{645}) - (4.68 \times A_{663}) = Chl\ b$$

$$(20.2 \times A_{645}) + (8.02 \times A_{663}) = Chl\ TOTAL$$

All electron transport data is based on the concentration of Chl.

Calibration of Oxygen Electrode

Fill the electrode chamber with well aerated distilled water and measure the temperature with the light on and off. There should be no difference; record the temperature. Measure the volume of the chamber.

Turn on the electrode and measure O_2 in water until a steady rate is indicated on the chart. This is the voltage of 100% O_2 saturation of X°C.

Add a pinch of sodium hydrosulfite and wait until the recorder comes to a

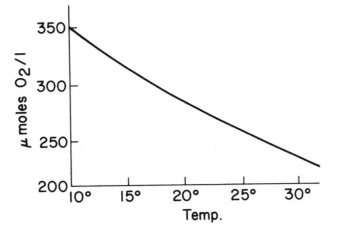

steady value. This is the voltage for zero oxygen. The difference between these two readings is the voltage of X umoles of O_2 per ml of the reaction chamber.

Calculate umoles O_2 per chamber per mV response.

Measurement of Photosynthetic Electron Transport

Make up the basic reaction mixture and add the required electron acceptors, electron donors, or inhibitors for the PSI-I+II, the PS-II, or the PS-I assay. Check that the pH is still near 8.0.

Add the reaction mix for the PS-II assay and 100 ug Chl of chloroplasts to the chamber. Measure O_2 until a steady rate is reached (this may require covering the chamber with foil or turning off room lights). When a steady trace is seen, turn on actinic light and mark the chart. Follow the trace until a linear response is clearly seen (30 to 120 sec.). Turn off actinic light, mark chart and allow recorder to continue until a steady dark rate is again seen. Proceed to the next reaction mixture or next sample of chloroplasts. If chloroplasts are limited, the PS-I+II assay can be measured, then DAD, Ascb, and DCMU are added to measure PS-I alone using the same chloroplasts.

Carefully wash out chamber after DCMU with several changes of water.

Measure the rate of O_2 evolution or uptake by measuring the mv change per minute on the chart and converting to a final value of umoles O_2 change per hour per mg chlorophyll.

If time is available, examine the effects of DCMU (inhibitor of electron transport) and methylamine (uncoupler of photophosphorylation) on mutant and normal green chloroplasts. At this point, you should be able to summarize the data for mutant and normal plants and compare the results of electron transport to the kinetics of fluorescence. You will now have the basic information on the activity of the photosynthetic light reaction and the relative rate of PS-I and PS-II.

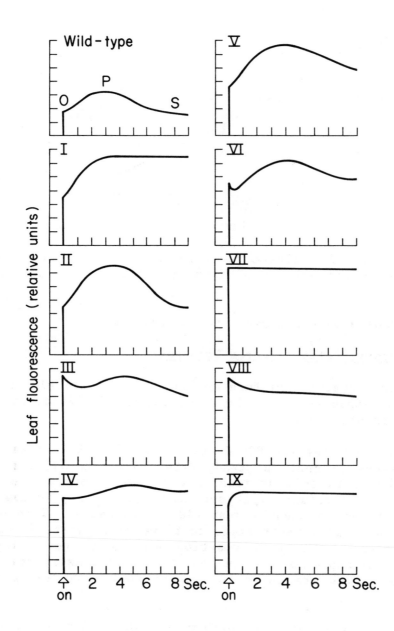

Analysis of Thylakoid Polypeptides

Chloroplast thylakoid membranes are isolated by grinding leaf material previously cut into one cm pieces. Cold grinding medium is used __without__ BSA. 50 to 10 grams fresh weight of washed leaves are ground in a cold blender for 15 to 30 sec. and filtered through Miracloth into cold centrifuge tubes. This chloroplast suspension is centrifuged briefly at 500 xg to remove cell debris and the chloroplast are pelleted at 5,000 xg for 10 min. The thylakoid membranes should be washed (to remove excess coupling factor) up to three times at 5,000 xg for 10 min in 10 mM Tricine-NaOH (pH 7.8) plus 10% sucrose to a chlorophyll concentration of about 0.5 mg/ml. At this point the sample may be stored at -70°C until needed.

__Electrophoresis of polypeptides__ employs a linear 10% acrylamide gel in a slab apparatus with 1.5 mm spacers and 8 mm wide wells. The following gel can be prepared:

Running gel, 10%	ml/30 ml

Note: Filter acrylamide and Tris with 0.45 ufilter, if possible.

30% Acrylamide stock	10 ml
(29% acrylamide, 0.8% bis-acrylamide)	
1.5M Tris, pH 8.8	7.5 ml
H_2O	12.5 ml
TEMED	10 ul
10% Persulfate (Add when ready to polymerize)	100 ul

Stacking gel, 5%	ml/10 ml

30% Acrylamide Stock	1.7
0.5M Tris, pH 6.8	2.5
H_2O	5.7
TEMED	5 ul
10% Persulfate (Add when ready to polymerize)	50 ul

The running gel is poured into the gel apparatus to within 5 mm of the well-forming comb. The top of the gel is leveled with a few drops of isoproponal carefully introduced with a pipette. The gel takes about 45-60 min. to polymerize, then the isoproponal is removed. The stacking gel is poured and the well-forming comb inserted. After 45-60 min. the comb can be removed and the well washed out and filled with upper running buffer.

Tank Buffer, upper	/600 ml

Glycine 0.19M	8.6 g
Tris 25 mM	1.8 g
LiDS 20%	3.0 ml
EDTA 200 mM	1.2 ml
H_2O	540 ml

Note: Filter LiDS and EDTA with 0.45 ufilter, if possible.

Tank Buffer, lower

Same as upper but without LiDS and EDTA

<u>Preparation of membrane samples</u>:

Membrane samples (fresh or thawed) are solubilized with a final conc. of 2% LiDS and 30 mM dithiothreitol. Samples are either maintained at 4°C or heated to 70°C for 4 min prior to loading on gels. Do one sample of each. Fifteen ul of chlorophyl is loaded, and the gels run at 4 watts constant power or 15 mA constant current. This should take 4 to 5 hr at 2°C in a 10 cm separations gel.

At the end of the electrophoresis remove the gel and stain for <u>chloroplast cytochromes</u>.

<u>TMBZ stain</u> (need 200 ml per gel) (3,3', 5,5' - Tetramethylbenzene)
0.1g TMBZ (final 0.5%)⎱
100 ml 100% Methonal ⎰ heat to dissolve
100 ml 2M Glacial acetic acid pH to 4.7 with NaOH

Incubate gel with above stain for 30 min in the dark. Then add 2 ml of 30% H_2O_2 to show up the peroxidase activity of the cytochrome. You may see three cytochromes in chloroplast polypeptides. This is the procedure of Thomas, et al., Anal. Biochem. 75, 168.

After staining for cytochromes the same gels are stained with Coomassie Brilliant Blue R-250 (0.2%; 50% methanol; 7% glacial acetic acid) for 1 or 2 hr; then destain with 50% methanol, 7% glacial acetic acid.

Analysis of chlorophyl-proteins from thylakoid membranes. This acylamid gel is prepared as before:

<u>Running gel 8%</u>	1.5 mm gel
30% acrylamide	8.0 ml
1.5M Tris, pH 8.8	7.0 ml
H_2O	15.0 ml
TEMED	7 ul
Persulfate	100 ul

<u>Stacking gel 4%</u>	
30% acrylamide	1.4 ml
0.5M Tris, pH 6.8	2.5 ml
H_2O	6.0 ml
TEMED	5 ul
Persulfate	50 ul

The sample thylakoid membrane preparation is treated with 30 mM octyl-glucoside in 2.0 mM Tris (brought to pH 7.0 with 1 M malate) with a ratio of 40 part octyl-glucoside to 1 part chlorophyl. To do this spin down a chl sample with 100 ug chl in a micro centrifuge tube and add 0.5 ml of the octyl-glucoside preparation. This is incubated 45 min at room temperature and 7 to 10 ul are loaded onto the gel per well. The electrophoretic separation for 1.5

hr at 2°C. The constant current was 15 mA.

<u>References</u>:

Delephelaire, P., N-H Chua, 1979 PNAS 75:111-115.

Camm, El, B.R. Green, 1980 Plant Physiol. 66:428.

Miles, D. 1982. Methods in Chloroplast Molecular Biology, Edelman, M. et al. (eds) pp. 75-108.

Thomas, P.E., Ryan, D. and Levine, W. 1976. Anal. Biochem. 75, 168-176.

Photosynthetic Mutants Materials

Plant Material:

    Three different nuclear mutants – Greenhouse flats to plant at least 50 plants/student, 800 seeds. Make second planting to be sure plants are at right age for screening. Plant 200 seeds of each mutant 14 days before experiment and 200 of each nine days before experiment. Grow in a warm greenhouse (28-32°C). Fertilize once a seek with liquid fertilizer high in nitrogen.

Screening for Photosynthetic Mutants:

    UV lamp, longwave:

        Spectronics B-100 longwave or
        UV Products UVL-56 longwave

    UV safety goggles:

        UV Products UVC 503 (or plastic goggles)

    Red plastic cut-off filter; any medium red plastic about 200 x 20 cm which will not pass the UV light.

    Toothpicks to mark plants.

    Darkroom to screen plants.

Leaf Fluorescence Kinetics:

    Filter fluorometer. Light source can be microscope lamp with blue plastic filter. Any type of chart recorder or storage oscilloscope can be used. I will provide two fluorometers but will need two recorders.

Isolation of Chloroplasts:

    Mortar and pestle (= 8 or 10 cm diameter)
    Ice buckets
    Ice
    Miracloth (20 cm square) or cheesecloth
    Funnel (= 10 cm diameter)
    Distilled water

Isolation medium – frozen (40 ml/isolation)	
Sucrose	0.4 M
NaCl	10 mM
$MgCl_2$	1 mM
$MnCl_2$	1 mM
Tricine	30 mM (pH 8.0)
EDTA	1 mM
B-Mercaptoethanol	5 mM
BSA	0.1 %

<u>Resuspending medium</u> - <u>frozen (20 ml/isolation)</u>
Sucrose   0.4 M
NaCl   10 mM
Tricine   20 mM (pH 8.0)

Spectrophotometer to measure Chl absorbance at 645 to 663 nm - Cuvettes (glass). Recording spectrophotometer would be useful but not required.

Oxygen electrode with recorder to measure oxygen evolved in photosynthesis and light source such as microscope illuminator and heat trap (0.2% $CuSO_4$).

I will provide two O_2 electrodes but will need two recorders, two magnetic stieres, and four microscope illuminators. Sodium hydrosulfite (1 g).

<u>Reaction Media</u> (basic part made up ahead of time and electron acceptor added later):

PS-II:	Tricine	40 mM (pH 8.0)
	NaCl	60 mM
	$MgCl_2$	4 mM
	PD	400 uM
	Potassium ferricyanide	2 mM
PS-I+II:	Tricine	40 mM
	NaCl	60 mM
	$MgCl_2$	4 mM
	Methyl viologen	100 uM
PS-I:	Tricine	40 mM
	NaCl	60 mM
	$MgCl_2$	4 mM
	Methyl viologen	100 uM
	DAD	500 uM
	Ascorbate	2.0 mM
	DCMU	8.0 uM

<u>Inhibitor</u> <u>and</u> <u>Uncoupler</u> (final concentration):

		Stock	
DCMU	8 uM	50 uM - 40% Mech	
Methylamine	10 mM	1 M	

<u>Make</u> <u>basic</u> <u>reaction</u> <u>mixture</u>:
        Tricine   ]
        NaCl   ]   pH 8.0
        $MgCl_2$

<u>Prepare</u> <u>stocks</u> <u>of</u> <u>reagents</u>:
        Potassium ferricyanide   100 mM
        Methyl viologen   1.0 mM

<u>Make</u> <u>up</u> <u>fresh,</u> <u>day</u> <u>of</u> <u>laboratory</u>:
        DAD (diaminodurene)   50 mM - 0.01N HCl

```
Ascorbate                              50 mM
PD (p-phenylenediamine)                50 mM - water
```

Chloroplast membrane proteins:

```
Cold blender = blender cup stored cold
Chloroplast Isolation buffer without BSA
10 mM Tricine/NaOH, 10 mM NaCl, pH 7.8
Slab electrophoresis equipment, 1.5 mm
30% Acrylamide stock
  (29.2% acrylamide, 0.8% bis-acrylamide)
1.5M Tris, pH 8.0
Isopropanal (25 ml)
Upper Tank Buffer
Lower Tank Buffer
2% LiDS (20% Stock) - or good SDS which will not
  precipitate at 4°C
30 mM Dithiothreitol (200 mM Stock)
3,3', 5,5' - Tetramethyl benzene 0.1 g/gel stained
Tethanol, 100%
Glacial acetic acid
NaOH
30% H₂O₂ 2 ml/gel stained

Coomassie Brilliant Blue R-250
Glycerol
30 mM Octyl glucoside in 2 mM Tris (pH to 7.0
  with 1 M malate)

Apparatus for electrophoresis
Gel trays for staining
Microcentrifuge
```

BEAN EMBRYO HORMONE IMMUNO-ELECTROPHORESIS

This procedure allows determination of proteins in crude extracts. The experiment examines the time course of disappearance of the seed lectin phytohemagglutinin (PHA) from the cotyledons and axis of bean (Phaseolus vulgaris) during early seedling growth.

Plant Material. Seeds were planted at 1 or 2 day intervals for 8 days prior to assay. Remove the axis or cotyledons from a plant, weigh, chop in a petri dish and homogenize in buffer (100 mg/ml) in a Polytron. After 10 min on ice, centrifuge in a Microfuge and use the supernatant in the assay at dilutions described below.

Day 0	axis and cotyledons	1/5, 1/10, 1/50, 1/100
Day 1	cotyledons	same as day 0
Day 2	cotyledons	same as day 0
Day 4	axis and cotyledons	undiluted, 1/2, 1/10, 1/50
Day 6	cotyledons	same as day 4
Day 8	cotyledons	

Prepare PHA standards in buffer.

Electrophoresis procedure

A. Preparing the agarose plates

1. Use 105x80 mm glass or Gel Bond (Bio-Rad) plates. Glass plates need to be washed in detergent, rinsed thoroughly in tap and distilled water, and dried. They should be in a 60°C oven prior to use.

2. Make a 1% agarose solution by placing 0.47 g of agarose powder (Bio-Rad, low M_r) in 47 ml of gel buffer in a 125 ml flask.

3. Heat the flask in a boiling water bath for 10-15 min. or until the agarose powder is completely dissolved.

4. Dispense 13 ml aliquots into 6x1 inch test tubes for storage, or place them into a 57°C waterbath.

   Note: The temperature at which the antibody is added is very critical. If it is too high, the antibody will be denatured, if it is too low, the gel will begin to polymerize.

5. Turn on the cooling system for the electrophoresis cell (Bio-Rad Model 1415) and add 1000 to 1200 ml electrode buffer to the tanks.

6. Add an appropriate amount of antibody (100 ul) to the warm agarose liquid, immediately cover with a piece of parafilm and mix gently by inverting the test tube. Avoid forming any air bubbles.

7. Place the gel plate on a leveled, warming table and pour the agarose onto the plate. Spread the agarose to the edges and corners of the

101

plate using the lip of the test tube. Remove the plates from the warming table and let them polymerize.

8. After the gel is polymerized, make wells with a #3 well puncher and the premade template. The wells should be 1.0-1.5 cm apart and 3 cm from the bottom edge, 8 wells per plate.

B. Applying the samples and running the gel

1. Dry the cooling stage of the electrophoresis unit and lay the plate onto it.

2. Use 3 layers of filter paper (Whatman #3, 20 x 10 cm) as bridges to connect the gel to the buffer.

3. Close the lid and turn on the power supply. Adjust the voltage to 100 V. The automatic ON/OFF switch on the electrophoresis unit allows one to apply samples without turning the power supply off and on.

4. Open the lid (the current should be turned off by the automatic switch) and apply 5 ul of sample to each well by using a Pipetman.

5. Close the lid, run the gel at 100 V for 18-24 hr.

C. Ending the run

1. Turn off the power supply and the cooling system.

2. Remove the gel from the unit and place onto a hard smooth surface.

3. Place 8 layers of filter paper (Whatman #3) on the gel and top with a thick glass plate (1/8") of the same size as the gel plate. Press the gel with 4 liters of water as weight.

4. After 15-30 min, remove the weight, glass plate and filter papers. Remove the last sheet of filter paper very carefully so as not to tear the gel.

5. Immerse the gel in a 0.1 M NaCl solution (1.46g/250 ml) for 10 to 15 min.

6. Wash (immerse) with d-H_2O for 10 min. Repeat.

7. Press the gel as described in Steps 3 and 4.

8. Dry the gel completely with a hair dryer.

    Note: When intending to stain for enzyme activiy, modify the procedures as follows:

    i.   Press the gel on the cooling stage with the cooling system running.

    ii.   Immerse the gel in NaCl in buffer (0.02 M K-phosphate, pH 6.8

102

containing 10^{-3}M DTT and 5×10^{-5}M $ZnCl_2$ for ADH staining), wash with buffer, keep the gel in cold.

   iii.   <u>Do not</u> dry the gel.

   iv.   Stain in the appropriate enzyme staining solution.

D. <u>Staining</u> <u>and</u> <u>destaining</u>

1. Immerse the dried gel plate in 150 ml staining solution for 15 to 20 min.

2. Pour off the staining solution (it is reusable after filtration). Rinse the gel with d-H_2O until the water is clear.

3. Pour destaining solution over the gel, swirl frequently but gently until the background becomes clear. Change the destaining solution if necessary.

4. Remove the gel and dry with a hair dryer.

E. <u>Calculations</u>

Using graph paper (or ruler), measure the height and width at the half-height of the rockets in mm. Multiply these two numbers and the result is the area under the peak in mm^2.

<u>Solutions</u>

A. <u>Monthony buffer</u>

Tricine	0.048 M	8.6 gm
Tris	0.162 M	19.62 gm
Calcium lactate	0.7 M	0.156 gm
NaN_3	6.2 M	0.403 gm
pH 8.7		
H_2O		

B. <u>Staining solution</u>

ETOH	450 ml
Acetic Acid	100 ml
Coomassie Brilliant Blue R	5 g
d-H_2O	450 ml

Dissolve Coomassie blue in ETOH and acetic acid; filter twice (Whatman #1) and add d-H_2O. Store on shelf in a dark container.

C. <u>Destaining solution</u>

ETOH	450 ml
Acetic Acid	100 ml
d-H_2O	450 ml

SUPPLEMENTS

SOURCES

Suppliers of odd items and chemicals.

Pectolyase Y23
Sigma Chemical

Cellulysin
Calbiochem

Purified Cellulase
Worthington

Large Forceps
7 inch, Straight Delicate
Roboz Surgical Instrument
1000 Connecticut Ave. NW
Washington, DC 20006

Small Forceps

Ted Pella Company

Plant Cons
Prepackaged Media
Linbro Detergent
Flow Laboratories
655 Old Springhouse Rd
McLean, Virginia 22102

GA7 Vessels
Magenta Corporation
4149 W. Montrose Ave.
Chicago, IL 60641

Sterile Hoods
Laminar Flow Inc
102 Richard Avenue
Ivyland, PA 18974

Sterile Hoods
NuAire Inc
2100 Fernbrook Lane
Plymouth, MN 55441

Growth Chambers
Percival
Boone, Iowa 50036

Coconut Water
Prepackaged Media
Gibco

MEDIA

This includes the standard CSHL plant cell culture media used during this course.

The following media should be autoclaved for 15 min. For agar media, add 10 gm of agar per liter (1%).

MS0 (shoot cultures, rooting)
500 ml H2O
100 ml MS Salts
1 ml vitamins
30 gm sucrose
fill to 1 liter with H2O
pH 6.0 with KOH

MS1 (tobacco medium)
500 ml H2O
100 ml MS Salts
1 ml vitamins
30 gm sucrose
1 ml 24D (0.5 mg)
fill to 1 liter with H2O
pH 6.0 with KOH

MS2 (BMS maize medium)
500 ml H2O
100 ml MS Salts
1 ml vitamins
20 gm sucrose
132 mg asparagine
4 ml 24D (2 mg)
fill to 1 liter with H2O
pH 6.0 with KOH

MS3 (tobacco medium)
500 ml H2O
100 ml MS Salts
1 ml vitamins
30 gm sucrose
6 ml IAA (3 mg)
0.6 ml Kinetin (0.3 mg)
fill to 1 liter with H2O
pH 6.0 with KOH

MS4 (tobacco regeneration)
500 ml H2O
100 ml MS Salts
1 ml vitamins
30 gm sucrose
0.6 ml IAA (0.3 mg)
20 ml DMAP (10 mg)
fill to 1 liter with H2O
pH 6.0 with KOH

MS7 (anther culture, rooting)
500 ml H2O
MS Salts
1 ml vitamins
30 gm sucrose
2 ml IAA (1 mg)
(3 gm active charcoal for anthers)
fill to 1 liter with H2O
pH 6.0 with KOH

Stocks

<u>20</u> <u>Liters</u> <u>MS</u> <u>Salts</u> <u>10X</u> <u>(Major</u> <u>and</u> <u>Minor</u> <u>Combined)</u>

Dissolve 330 gm NH4NO3 in 4 liters H2O, add to carboy
    "   380 gm KNO3   "   "   "   "   "   "
    "   34 gm KH2PO4   "   "   "   "   "   "
    "   88 gm CaCl2-2H2O   "   "   "   "   "
    "   74 gm MgSO4-7H2O   "   "   "   "   "
Remove 2 liters from this 20 liter mix, add the following:
    7.5 gm Na2EDTA
    5.6 gm FeSO4
    1.2 gm H3BO3
    3.4 gm MnSO4-7H2O
    2.1 gm ZnSO4-7H2O
    166 mg KI
    50 mg Na2MoO4-2H2O
    5 mg CuSO4-5H2O
    5 mg CoCl2-6H2O
Add these 2 liters back to carboy. Mix up by draining several liters into a flask, and then add back to the carboy. Repeat several times. The carboy should be cleaned before each batch, but do not use bleach or other potentially toxic substance.

<u>Hormones</u>

Dissolve 50 mg in 97 ml H2O plus 3 ml 1 M KOH. Use gentle heat if necessary to dissolve the hormone. Filter sterilize and add to a sterile bottle. Store in refrigerator. These stocks can be pipetted outside the hood, in the kitchen, provided the media is to be autoclaved.
    24D = 2,4-dichlorophenoxyacetic acid
    IAA = Indoleacetic acid
    NAA = Naphthalene acetic acid
    DMAP = Dimethylallylaminopurine
    BAP = Benzylaminopurine
    Kinetin

<u>Vitamins</u>

These vitamins are the B5 vitamins of Gamborg, which we prefer.
1 gm Thiamine HCl
100 mg Pyridoxine HCl
100 mg Nicotinic Acid
10 gm myo-Inositol
95 ml H2O
Heat stir to dissolve. Filter sterilize, then aliquot as 1 ml fractions to be frozen. Use 1 tube (1 ml) per liter media.

References
T. Murashige, F. Skoog 1962 Physiologia plantarum 15:473-497.
O. Gamborg, L. Wetter 1975 Plant tissue culture methods (National Research Council of Canada)

Tobacco Protoplast Media

Amount per liter of medium, final

Salts	NH	5XH	HHLS	LH
$CaCl_2 \cdot 2H_2O$	750 mg	750 mg	750 mg	750 mg
$CoCl_2 \cdot 6H_2O$	.025	.025	.025	.025
$CuSO_4 \cdot 5H_2O$	.025	.025	.025	.025
FeNaEDTA	40	40	40	40
H_3BO_3	3	3	3	3
KI	.75	.75	.75	.75
KNO_3	2500	2500	2500	2500
$MgSO_4 \cdot 7H_2O$	250	250	250	250
$MnSO_4 \cdot H_2O$	10	10	10	10
$Na_2MoO_4 \cdot 2H_2O$	.25	.25	.25	.25
$NaH_2PO_4 \cdot H_2O$	150	150	150	150
$(NH_4)_2SO_4$	134	134	134	134
$ZnSO_4 \cdot 7H_2O$	2	2	2	2

Organic Constituents

	NH	5XH	HHLS	LH
myo-Inositol	0.1 gm	0.1 gm	0.1 gm	0.1 gm
Sucrose	137 gm	137 gm	70 gm	70 gm
Xylose	0.25 gm	0.25 gm	0.25 gm	0.25 gm
Nicotinic acid	1 mg	1 mg	1 mg	1 mg
Pyridoxine-HCl	1	1	1	1
Thiamine-HCl	10	10	10	10
Benzylaminopurine	0	5 mg	1 mg	.2 mg
Naphthalene Acetic Acid	0	15	3	.2
pH (before autoclaving)	5.8	5.8	5.8	5.8

Maize Cell Culture Media

Chu, C.C. et al., 1975. Establishment of an efficient medium for anther culture of rice through comparative experiments on the nitrogen sources. Scientia Sinica 18(5):659-668. (casein hydrolysate not used in original article)

N6 media

I. <u>Salts</u>

<u>Major elements</u>	<u>mg/1</u>	<u>mM</u>
$(NH_4)_2SO_4$	463	3.5
KNO_3	2830	28.0
$CaCl_2*2H_2O$	166	1.13
$MgSO_4*7H_2O$	185	0.75
KH_2PO_4	400	2.94
Na_2*EDTA	37.3	0.20 (Na)
$FeSO_4*7H_2O$	27.8	0.10 (Fe)

<u>Minor elements</u>	<u>mg/1</u>	<u>micro-M</u>
H_3BO_3	1.6	25.8
$MnSO_4*1H_2O$	3.3	19.5
$ZnSO_4*7H_2O$	1.5	5.2
KI	0.8	5.0

II. <u>Organic constituents</u>

	<u>mg/1</u>
thiamine HCl	1.0
glycine	2.0
pyridoxine	0.5
nicotinic acid	0.5
casein hydrolysate	100
sucrose	20 g/1

III. <u>pH</u> Adjusted to 5.8 with NaOH (and HCl, if necessary).

Maize Regeneration Media (PRM Media)

Core Media

MS Stock Salts	100 ml/l
2,4-D	2
coconut water	20
organic acids	10
PRM vitamins	10
sugar stock	10 ml/l
mannitol	54.7 gm/l
sucrose	20 gm/l
casein hydrolyzate	200 mg/l
asparagine	150 mg/l
glucose	250 mg/l

Organic Acids 100x pH 5.5

sodium pyruvate	500 mg/l
citric acid	1 gm/l
malic acid	1 gm/l
fumaric acid	1 gm/l

PRM Vitamins

nicotinamide	100 mg/l
pyridoxine HCl	100
calcium pantothenate	50
folic acid	20
p-aminobenzoic acid	1
biotin	.5
choline chloride	50
riboflavin	10
ascorbic acid	100
vitamin B12	1

Sugar Stock

fructose	12.5 gm/1
ribose	12.5
xylose	12.5
mannose	12.5
cellobiose	12.5

All stock solutions are filter sterilized.   The pH of PRM is adjusted to 5.8.

NOTES ON PLANT CELL CULTURE TECHNIQUES

These notes and suggestions were written by Maureen Hanson (University of Virginia), for the 1981 CSHL plant course, and revised in 1984.

Sterile Air Hoods

The chief items of equipment important for plant cell and tissue culture which are not typical components of a molecular biological laboratory are horizontal laminal flow hoods and an inverted microscope. Biohazard hoods can, of course, be used for plant culture work. However, since plant cultures are usually not hazardous, a horizontal air flow hood, which is easier to install and less expensive, is completely satisfactory.

Besides the routine factors to plan for when purchasing a hood such as size fit through doorways and in the room, availability of service and company reputation, and weight of hood (typically ranges from 400-900 lbs), noise level is an important aspect to consider.

Sterile air hoods can be purchased in different table and total heights and lengths. A 6-foot hood is the smallest size in which 2 people can work comfortably. Unfortunately, the larger the hood, the noisier it is in operation; noise levels over 70 decibels are uncomfortable and may cause hearing damage. One 6-foot hood that produces less than 70 decibels of noise is produced by Laminar Flow Inc. This hood is quieter than most because it is made of formica-covered wood, has a sound-muffling baffle in the front, and contains two air intake fans (most hoods contain only one).

Most companies have slightly different hood designs which have a patented feature affecting filter placement of air flow. Baker Edgeguard hoods have a special patented air flow system (intake on edges of the working space) which reduces operator disturbance of the sterile area.

If several hoods are to be placed in one room, company representatives should be consulted regarding the noise level. Hood certification companies, whose personnel have experience with products of many different companies, are also a good source of information.

The best location for a sterile air hood is in a separate room used only for eukaryotic culture work. A corner of a large laboratory, away from traffic, may also be suitable. Another advantage of a separate room for the hood(s) is extension of hood filter life by minimizing particle content of the air. A prefilter (such as is used in hospital operating rooms) on the building air system is particularly useful in a sterile air hood room, and in some cases may be necessary to reduce the particle count to the point where the hood is efficient. Air hoods do not remove all particles from air, merely a high percentage of them.

Environments for Plant Tissue Culture

Either growth chambers or lighted growth rooms can provide a suitable environment, usually between 24 and 28°C with a day-night cycle of 200-400 ft

112

candles of light. Cool white or warm white fluorescent fixtures are most often used. Unfortunately, the exact light conditions for plant cultures are often omitted from materials and methods, or foot-candles reported without the type of light specified. Drying of plates and cultures can be a problem over the long culture periods necessary. Either humidified growth chambers or humidified vessels (plastic refrigerator boxes containing open dishes of distilled water) can solve the dehydration problem. Note that chambers and rooms should be relamped frequently for uniformity of growth conditions.

Glassware Handling

1. Reused tissue culture glassware must be carefully cleaned. Residues harmful to bacterial, algal, or yeast cultures can severely affect higher plant (and animal) cultures.

2. Alkaline detergents used in many microbiological kitchen's dishwashers can leave harmful residues. Also, salts and inhibitory compounds can accumulate in dishwashers used for microbial and biochemical glassware. For best results, if glassware is to be machine-washed, a dishwasher designated for animal and plant culture work may be necessary. Non-toxic tissue culture detergents (e.g., 7X from Linbro, Flow Labs) are available in forms for either hand- or machine-washing of glassware. Handwashed glassware should be rinsed thoroughly, 6 times with tap water and 5 times with distilled water.

3. Sterile plastic disposable pipets are convenient for manipulating cell cultures. Glass pipets used for transferring suspension cultures are particularly difficult to clean. The following procedure has proved successful: Pipets are placed in pipet jars with a Wescodyne solution immediately after use. Then the pipets are washed with alkaline detergents in a microbiological dishwasher. This is followed by 6-24 hours in a pipet jar containing 7X detergent. Pipets are rinsed overnight with tap water followed by at least 5 exchanges with distilled water in a Nalgene pipet-washer.

4. Tissue culture pipets must be plugged with cotton, which must be removed with forceps before washing. Bellco manufactures an efficient automatic pipet plugger.

Water

In these protocols "water" is double distilled. While ordinary distilled water may be suitable for explant cultures, double (glass) distilled water should be used in all culture media to eliminate a source of variability.

Plant Tissue Culture Media

Components of plant tissue culture media can be divided into these categories: salts, vitamins, hormones, sugars, extra organic compounds, and gelling agent. For ease of bookkeeping, a master "media notebook" in which all plant media formulations used in the lab are listed, is useful. We number our media consecutively using a prefix symbolizing the salt formulation (for example, MS413 is the 413th medium containing MS salts our lab has used). Every individual experimenting with new formulations in the lab uses the master number series and records each medium variation in the master list. The medium formulations are always listed as follows:

    salts
    vitamins
    hormones
    sugars
    extra organic compounds
    gelling agent
    method of pH adjustment
    sterilization method

In order to reduce chance of error, ingredients are always added in this order. Stock solutions can be conveniently identified by using a different color tape for each class of medium component. Each stock solution is identified by name, date made, and initials of the person who made the stock.

Salts:
A number of salt formulations and media are available as commercial premixes, due to their use in the micropropagation industry. MS salt mix is available at reasonable cost from both GIBCO and FLOW laboratories. The premix price of the other salt formulations encourages laboratories to make up their own salt solutions from concentrated stocks. Stock salt solutions can often be autoclaved and stored at room or refrigerator temperature. The iron stock tends to become contaminated easily if not handled with sterile techniques.

Vitamins:
The most commonly used components of plant tissue culture media which are referred to as "vitamins" are myo-inositol, glycine, thiamine, pryridoxin, and nicotinic acid. Particularly in protoplast culture media some additional ones--such as biotin and folic acid--have been found to be useful. Myo-inositol and thiamine have been shown to be essential in a number of cultures; often the other vitamins are actually not required. These can be stored refrigerated for up to a month, or it may be convenient to mix the stocks for 1 liter together, and store them as frozen aliquots.

Hormones:
Hormone stock solutions can be made up in water, acid, or base (see table 2) and stored in the refrigerator. Auxins, cytokinins, and gibberellins are the hormones most often used in culture media. Certain hormones are not stable on storage (especially IAA) and stock solutions of these should be made up fresh. For maximum reproducibility, unstable hormones be filter-sterilized and not autoclaved. In order to reproduce published media, the sterilization method reported should be followed. Unfortunately, many materials and methods sections in the literature omit the method of media sterilization, which can critically affect the hormone concentration.

Sugars:
Sucrose is the sugar used most often as an energy source, although glucose is not an uncommon component. Sugars are also used as osmotic agents in protoplast culture media; non-metabolized sugars such as mannitol and sorbitol are often included for this purpose. Sugar solutions often brown or yellow upon autoclaving, sometimes having harmful effects on the cultures. This problem is avoided for delicate cultures by filter-sterilizing or autoclaving

the sugars separately from the salts. (see also Method of Sterilization).

Extra organic compounds:
Additional amino acids, organic acids, and undefined organic mixtures such as coconut milk, casein hydrolysate, or conditioned medium are sometimes included.

Gelling agent:
Difco bacto-agar has been the most commonly used gelling agent for plant tissue cultures until recently. It is suitable for many applications, particularly with cooperative species and less demanding techniques, such as explant culture. Some investigators find removing yellow filtrates by washing agar multiple times with distilled water, and then ethanol, to be necessary. A recent alternative is Phytagar (Gibco), which appears to be cleaner than Bacto-agar and has been selected by the supplier for plant tissue culture suitability. In addition, agar substitutes (Plantagar by Flow Labs) are available. For protoplast culture, the low melting temperature agaroses (Sigma, FMC), have often produced good results due to their high purity and the low temperature at which solutions can be added to cells in liquid suspension.

Method of pH Adjustment
A pH between 5.4 and 5.8 is generally desirable for plant culture. Adjusting unbuffered media to 6.0 with KOH usually results in a pH of 5.6 to 5.8 after autoclaving. Most plant media are not deliberately buffered, although MES has been used for this purpose. Certain components such as casein hydrolysate will act as buffers; filter-sterilized media or media including such components should be adjusted to 5.6 or 5.8 instead of 6.0. In cases where pH is critical, media should be checked after autoclaving. Indicator dyes such as chlorophenol red and bromocresol purple have been found non-toxic to those cultures tested.

Method of Sterilization
Especially when easily-culturable species are in use, most labs autoclave the complete medium, or the complete medium except for a filter-sterilized hormone. For reproducibility, the medium should always be autoclaved for the same length of time (preferably with the same autoclave) and removed immediately. Autoclaving times necessary will vary between 15 and 25 min depending on the individual autoclave. Small precipitates are not uncommon in MS media after autoclaving. Filter-sterilization of the complete medium (except agar) is often the method of choice when culturing recalcitrant species and protoplasts. The necessity of filter-sterilized media for proper growth of some cells can sometimes be avoided by autoclaving sugars separately from the other components.

Storage of medium
Medium will last longest if poured petri dishes are dried with cover closed in the hood for a few hours and then placed back right-side-up in the petri dish sleeves while still in the hood. These sleeves, closed tightly with a twist-tie, can be conveniently stored in an empty petri dish box in numerical order. We have had sleeves remain sterile and hydrated for up to a year at room temperature when this procedure is followed, and have successfully performed explant culture with media after one year at room temperature. However, the shelf life of media is under debate and varies with media components. For example, media containing IAA or antibiotics should not be

stored for long time periods. In general, we prefer to use media not stored more than three months. Liquid media in bottles can be stored in the refrigerator. Because condensation often results in contamination in dishes stored in sleeves in the refrigerator, we store plates at room temperature. Dishes should be stored in the dark, especially those containing IAA. Medium should be labelled with date made for recording when used in experiments.

Vessels

Plastic petri dishes (9 cm diameter, 20 or 25 mm depth) are often used for callus and explant culture. A variety of test tubes, vials, smaller petri dishes, multiwell dishes (Costar, Falcon) and glass petri dishes are also appropriate for particular applications. The need for larger vessels to accomodate regenerated plants has stimulated the development of additional containers, including the disposable "Plantcons" (Flow) and reusable GA7 vessels (Magenta).

Standard Plant Tissue Culture References

Barz, W., Reinhard, E., Zenk, M. eds. 1977. Plant tissue culture and its biotechnological application. Springer-Verlag, NY.

Butcher, D.N. and D.S. Ingram. 1976. Plant tissue culture. Edward Arnold Ltd., London.

Chaleff, R.S. 1981. Genetics of higher plants: applications of cell culture. Cambridge Univ. Press, N.Y.

Cocking, E., Peberdy, J., eds. 1974. Propagation of higher plants through tissue culture, a bridge between research and application. National Technical Information Service, U.S. Dept. Commerce, Springfield, VA 22161.

Conger, B. 1981. Cloning agricultural plants via in vitro techniques. CRC Press, Boca Raton, Florida.

Dodds, H., Roberts, L. eds., 1982. Experiments in plant tissue culture. Cambridge University Press, Cambridge.

Dudits, D., Farkas, F., Maliga, P. eds. 1976. Cell genetics in higher plants. Hungarian Academy of Sciences, Budapest.

Durbin, R.D., ed. 1979. Nicotiana: Procedures for experimental use. U.S. Department of Agriculture, Technical Bulletin 1586. Available for $7.25 in advance from Superintendent of Documents, Washington, D.C.

Earle, E., Demarly, Y. eds. 1982. Variability in plants regenerated from tissue culture. Praeger, NY.

Ferenczy, L., Farkas, L., Lazar, G. eds. 1980. Advances in protoplast research. Proc. Fifth International Protoplast Symposium, Szeged. Pergamon Press, Oxford.

Fujiwara, A. ed. 1982 Proc. in plant tissue culture 1982. Fifth Intl. Cong. Plant Tissue and Cell Culture. Japanese Assoc. for Plant Tissue Culture,

Tokyo.

Gamborg, O.L., Wetter, L.R. eds. 1975. Plant Tissue Culture Methods. Published by the NRC of Canada, Prairie Regional Lab, Saskatoon, Saskatchewan S7N OW9, Canada. Paperback. Available from N.R.C., C.N.R.C., Publications, Ottawa, Ontario, Canada K1A OR6.

Hughes, K., Henke, R., Constantin, M. 1978. Propagation of higher plants through tissue culture, a bridge between research and application. National Technical Information Service, U.S. Dept. Commerce, Springfield, VA 22161

Potrykus, I., Harms, C., Hinnen, A., Hutter, R., King, P., Shillito, R. 1983. Protoplasts 1983, lecture proceedings. Sixth international symposium. Basel, Birkhauser-Verlag, Basel.

Reinert, J. and Y.P.S. Bajaj. 1977. Applied and fundamental aspects of plant cell, tissue, and organculture. Springer, N.Y.

Reinert, J., Yeoman, M. 1982. Plant cell and tissue culture, a laboratory manual. Springer Verlag, Berlin.

Sen, S., Giles, K. 1983. Plant cell culture in crop improvement. Plenum press, NY.

Street, H.E., ed. 1979. Plant tissue and cell culture (2nd Edition) Botanical Monographs, Vol. II. U. Cal. Press, Berkeley.

Thorpe, T.A., ed. 1978. Frontiers of plant tissue culture. Intern. Assoc. Plant Tissue Culture, Calgary, Canada.

Thorpe, T.A., ed. 1981. Plant tissue culture, methods and applications to agriculture. Academic Press, NY.

Vasil, I. ed. 1980. Perspectives in plant cell and tissue culture. Int. rev. cytology, supp 11B. Academic Press, NY

White, P.R. 1963. The cultivation of animal and plant cells. 2nd ed. Ronald Press, N.Y.

Wilmer, E.N. ed. 1966. Cells and tissues in culture. Vol. 3. Academic Press, N.Y.

NOTES ON PLANT REGENERATION FROM CULTURES OF MAIZE

C.E. Green and C.A. Rhodes
Department of Agronomy and Plant Genetics
University of Minnesota
St. Paul, MN 55108

Reprinted from Maize for Biological Research, W.F. Sheridan, ed. Copyright
1982, Plant Molecular Biology Association, PO Box 5126, Charlottesville, VA
22905.

Tissue cultures capable of plant regeneration may be efficiently initiated
from immature embryos of maize and more recently from several other sporophytic
tissues including immature tassels and ears (Green and Phillips, 1975; Freeling
et al., 1977; Green, 1977; Molnar et al., 1980; Rhodes et al., 1982). The
morphology of these cultures is typically one of a complex association of shoot
and root meristems as well as less differentiated tissues (Springer et al.,
1979). This morphology has been diagnostic of the capacity to regenerate
plants of maize with the exception of a new culture type to be discussed below.

The various sporophytic tissues which can be used to initiate regenerable
tissue cultures have their individual experimental advantages. Immature
embryos are available in large numbers from a single pollination and they
develop in relative synchrony on the same ear. Consequently they can be used
effectively to compare genetic differences or experimental treatments among
sibs. Immature tassels, on the other hand, are obtained from donor plants
which are usually 6 to 7 weeks old. This allows time for the expression of
certain genetic markers which permit the identification of cytogenetically
altered plants, such as haploids, monosomics, or plants with deletions, prior
to initiation of cultures.

Tissue culture Initiation

To initiate cultures from immature embryos, the husked intact ear is
sterilized in 2.5% sodium hypochlorite plus a small amount of detergent for 20
min and then rinsed three times in sterile water. Each rinse should be for a
minimum of 5 min. Using sterile instruments in a sterile environment, such as
a laminar air flow hood, the crown of each kernel is cut off and a rounded-tip
spatula is used to scoop out the endosperm. Embryos up to 2 mm in length
usually remain associated with the endosperm and can be seen with the unaided
eye and removed from the endosperm with the tip of the spatula. The embryo is
placed on the culture medium with the mebryonic axis against the medium. This
embryo orientation permits extensive proliferation in the scutellum and
minimizes germination and further development of the embryonic axis. Up to ten
embryos fit easily in a 100 x 25 mm Petri dish containing 50 ml of medium.
Masking tape or Parafilm strips provide a good seal for the dishes.

Immature tassels for tissue culture studies are typically obtained from 6
to 7-week-old field or greenhouse grown plants. The tassels in these plants
are usually 2 to 3 cm in length and are encased in many layers of young leaves.
Sterilization of the tassel is not necessary if the leaves surrounding it are
carefully removed in a sterile environment. The entire tassel is cut into 1 to
2 mm thick cross sections and all these pieces are immediately transferred to a
Petri dish containing 50 ml of culture medium to minimize desiccation.

Orientation of the pieces on the medium does not appear to be critical.

Although several growth medium formulations will support the growth of regenerable tissue cultures of maize, the most frequently used media are MS (Murashige and Skoog, 1962) and N6 (Chu et al., 1975; Chu, 1978). A particularly critical ingredient in the medium is the auxin which must be of sufficient potency and concentration to suppress germination of immature embryos and root formation in immature tassel sections during the initiation of the tissue cultures. 2,4-dichlorophenoxy acetic acid (2,4-D) at 0.5 to 1 mg/l has been used with the greatest success as an auxin for tissue culture propagation in maize.

After three to four weeks of incubation at 25-28°C under low fluorescent light (1.5 W/m^2), regenerable cultures from immature embryos or tassels can be identified by the presence of scutellar-like bodies. Vigorous cultures will be light green or yellow and may have small leaves developing among the scutellar-like bodies. The growth and regenerability of these cultures can be maintained for several years by transferring pieces about 5 mm in diameter to fresh medium every 3 to 4 weeks.

Plant Regeneration

To regenerate plants from these cultures the auxin (2,4-D) must be greatly lowered or omitted from the medium. Depending on the degree of organization in the culture it may be useful to lower the auxin in steps via successive transfers or to remove it entirely in one transfer. Shoots which emerge are transferred to 250 ml Erlenmeyer flasks containing 75 ml of medium without 2,4-D and incubated under bright fluorescent light (8-10 W/m^2; 16/8 hr photoperiod). When an adequate root system has developed, the plant is transferred to 2 inch plastic pots containing a 1:1 mixture of sterile soil and vermiculite. It is critical at this stage to thoroughly rinse all the growth medium off the roots before transplanting the plant into soil. The plants should be kept in a humid, brightly lighted environment for 2 to 4 weeks until established, before transplanting to a greenhouse or field. Regenerated plants often have more vigor and are more likely to produce morphologically normal tassels and ears if the various steps in regenertion and establishment are carried out in as short a time as possible. Pollination of regenerated plants is accomplished by selfing or crossing with other plants to produce the desired kernels and progeny. Seeds from regenerated plants are generally grown into plants using standard cultivation practices.

Genetic Analysis

Cytogenetic analysis of plants regenerated from diploid maize cultures, eight-months old or less, has indicated a high degree of chromosomal stability (Green et al., 1977; McCoy and Phillips, 1982). The instability which does occur usually involves missing or broken chromosomes. Abnormal plants may have sectored tassels such that one or more of the tassel branches exhibit partial pollen sterility and one or more of the branches have normal fertility. Meiotic analysis of pollen mother cells in young tassels has identified a total of 9 cytologically abnormal plants among 277 examined (Green et al., 1977; Edallo et al., 1981; McCoy and Phillips, 1982). The limited data available on the chromosome constitution of plants regenerated from older tissue cultures indicates a much higher frequency of abnormalities. Eleven plants regenerated from three-year old cultures all possessed the same phenotype (i.e. oppositely arranged leaves and ca. 100 cm tall) at maturity (Green et al., 1977). Three plants were analyzed cytologically and each plant contained a broken chromosome 6, deficient for the distal third of the long arm. This is an indication that cytological abnormalities may be increased in older cultures.

An interesting aspect of recent studies is that tissue culturing induces considerable variation in the genetic makeup of cells and that this variability is recovered in the progeny of regenerated plants. A high frequency of spontaneous mutations with simple Mendelian inheritance has been observed in R2 progenies (Edallo et al., 1981). The type of endosperm and seedling mutants found were similar to spontaneous mutants described in maize. A high rate of spontaneous mutation makes these cultures of particular interest in selection experiments.

Mutants have been selected from maize tissue cultures in two separate experiments. Selection for resistance to H. maydis race T pathotoxin in diploid, Texas male sterile cultures resulted in the recovery of resistance which was transmitted to regenerated plants and progeny (Gengenback et al., 1977). This selection also changed male sterility to fertility. Both resistance to the disease and fertility were inherited as maternal traits. Selection for resistance to lysine plus threonine in diploid cultures also produced resistance which was recovered in the pogeny of regenerated plants (Hibberd and Green, 1982). This lysine plus threonine mutant, Ltr*-19, overproduces threonine in large quantities in the seed and is inherited as a dominant gene.

Somatic Embryogenesis

A major new type of maize tissue culture has been developed recently which is distinguished from those discussed previously by its friability, rapid growth, and capacity to regenerate plants by somatic embryogenesis (Green, 1982). These cultures are light yellow and appear undifferentiated in that little organization is visible to the eye. Closer examination by light and scanning electron microscopy, however, reveals somatic embryos which follow very closely the known developmental sequence for zygotic embryos (Randolph, 1936).

These embryogenic cultures are initiated directly from the scutellum of immature embryos or as spontaneous sectors growing from established organogenic tissue cultures of the type discussed earlier in this paper. They can be initiated and maintained on both MS and N6 medium containing 0.5 to 1.0 mg/l 2,4-D. the cultures are first noticed as small friable regions which are light yellow. These cultures may not show visible organization initially but frequently within 2 weeks of their appearance, microscopic examination reveals embryos at globular or slightly more advanced stages of development. Once established these cultures exhibit a rapid growth rate and must be subcultured every 2 weeks. They are maintained by transferring 4 to 5 pieces of callus, each 5 to 10 mm in diameter, to 100 x 25 mm Petri dishes containing 50 ml of fresh medium.

The established cultures exhibit a high degree of embryogenic activity; frequently hundreds of embryos per callus are visible. They are first seen as small globular structures and their continued development is evidenced by differentiation of the suspensor, scutellum and embryonic axis tissues. Embryo development up to the coleoptilar stage (Abbe and Stein, 1954) progresses quite normally on either MS or N6 media containing 2% sucrose and 0.5 to 1.0 mg/l 2,4-D. Development beyond this stage is very abnormal unless the callus is transferred to medium lacking 2,4-D but with 6% sucrose to increase the osmolarity of the medium. Maturation proceeds rapidly on this medium and after 10 to 14 days embryos have developed which are similar in size to those found in seeds. Their most prominent features are a well-developed scutellum and embryonic axis. These embryos germinate rapidly when transferred to medium lacking hormones and 2% sucrose. Shoot and root development occurs simultaneously and the young plants are then grown under conditions described in the Plant Regeneration section.

These friable embryogenic callus cultures are appropriate for the initiation of suspension cultures which are grown in liquid MS medium containing 2,4-D and aerated on a gyrotory shaker. The resulting cultures are well-dispersed and composed of cell aggregates ranging from about 2 mm in diameter down to single cells. Embryo development rarely proceeds as far as the globular stage in these cultures. When these suspensions are plated on agar-solidified MS medium, active embryogenesis is observed in the resulting callus cultures.

Other Important Factors

The genotype of the donor tissue substantially influences the ease with which tissue cultures of maize are obtained and the duration of regenerability. When using immature embryos, some genotypes (i.e. A188, WF9, and ND203) produce regenerable cultures very efficiently while similar cultures are initiated with great difficulty from other genotypes, such as W23 and A632. Regenerable cultures have been initiated from immature embryos of 70% of the approximately 40 different genotypes we have examined, without modification of the standard cultural regime. Modifications in the growth medium have improved the performance of some more difficult genotypes. Typical modifications include adjustments in the 2,4-D concentration and alternative macro- and micro-nutrient formulations of the medium. Characteristic differences among genotypes are also noted when cultures are initiated from immature tassels. Growth conditions can and should be optimized to produce the most desirable cultures from specific genotypes.

The developmental stage of the donor tissue is an important factor in the initiation of regenerable tissue cultures. As immature embryos become more fully developed, they rapidly lose the ability to initiate regenerable cultures (Green, 1977). Embryos 3 mm or longer are generally ineffective for culture initiation. For most genotypes, embryos which are 1 to 2 mm long most reliably produce the desired cultures. Immature tassels more frequently produce regenerable cultures when they are 1 to 3 cm long.

Incubation conditions used for maize tissue cultures are very similar to those for cultures of other species. The growth rate of maize cultures increases with temperature up to 30°C. The most frequently used temperature for incubation is 28°C. Temperature control in growth rooms and chambers should be as uniform as possible to minimize the condensation of water from the growth medium onto the lids and walls of Petri dishes and other containers. The greater the fluctuation in temperature, the greater the problem of

121

condensation. Severe condensation can interfere with culture growth as well as increase the chances of microbial contamination. Humidity control in growth rooms and incubators is generally not necessary if dishes and flasks are sealed properly with tape, parafilm, or aluminum foil. These seals are also important in preventing microbial contamination of the cultures.

Although not dependent on light, growth of the cultures is often improved and regeneration yields more vigorous plants when they are grown in the light. Typical light sources include standard Cool-White, Grow-Lux, or Agro-Lite fluorescent fixtures. These provide light intensities up to 1.5 W/m^2 which is more than adequate for the cultures. Typical photoperiods are 12-16 hr of light per day.

Cultures should be subcultured as frequently as necessary to maintain vigorous growth. The actual time interval is influenced by the type of culture, volume of culture medium, and the amount of inoculum used for the subculture. Cultures with rapid growth rates, such as the friable embryogenic lines, are transferred every two weeks while slower growing cultures are transferred about every four weeks. Careful management of the subculture and incubation conditions is especially important for the maintenance of the regenerability of cultures. The potential to regenerate numerous plants from tissue cultures of maize can be maintained for several years if necessary.

References

Abbe, E.C. and O.L. Stein. 1954. The growth of the shoot apex in embryogency. Amer. J. Bot. 41:285-293.

Chu, C.C., C.C. Wang, C.S. Sun, C. Hsu, K.C. Yin, C.Y. Chu. 1975. Establishment of an efficient medium for anther culture of rice through comparative experiments on the nitrogen sources. Sci. Sinica 18:659-668.

Chu, C.C. 1978. The N6 medium and its application to anther culture of cereal crops. In: _Proceedings of Symposium on Plant Tissue Culture_. pp. 43-50. Science Press. Peking.

Edallo, S., C. Zucchinal, M. Perenzin, and F. Salamini. 1981. Chromosomal variation and frequency of spontaneous mutation associated with in vitro culture and plant regeneration in maize. Maydica 26:39-56.

Freeling, M., J.C. Woodman, and D.S.K. Cheng. 1977. Developmental potentials of maize tissue cultures. Maydica 21:97-112.

Gengenback, B.G., C.E. Green, and C.M. Donovan. 1977. Inheritance of selected pathotoxin resistance in maize plants regenerated from cell cultures. Proc. Natl. Acad. Sci. U.S. 74:5113-5117.

Green, C.E. 1977. Prospects for crop improvement in the field of cell culture. Hort. Science 12:131-134.

Green, C.E. 1978. In vitro plant regeneration in cereals and grasses. In: _Frontiers of Plant Tissue Culture_. pp. 411-418. T.A. Thorpe, ed., Univ. of Calgary, Calgary, Alberta.

Green, C.E. 1982. Somatic embryogenesis and plant regeneration from friable callus cultures of maize. (in preparation).

Green, C.E. and R.L. Phillips. 1975. Plant regeneration from tissue cultures of maize. Crop Sci. 15:417-421.

Green, C.E., R.L. Phillips, and A.S. Wang. 1977. Cytological analysis of plants regenerated from maize tissue cultures. Maize Genet. Coop. News Lett. 51:53-54.

Hibberd, K.A. and C.E. Green. 1982. Inheritance and expression of lysine plus threonine resistance selected in maize tissue culture. Proc. Natl. Acad. Sci. U.S. 79:559-563.

McCoy, T.J. and R.L. Phillips. 1982. Chromosome stability in maize (Zea mays L.) tissue cultures and sectoring among regenerated plants. Can. J. Genet. Cytol. in press.

Molnar, S.J., P.N. Gordon, and T.B. Rice. 1980. Initiation of totipotent tissue cultures from undeveloped axillary and secondary ears. Maize Genet. Coop. News Lett. 54:52-53.

Murashige, T. and F. Skoog. 1962. A revised medium for rapid growth and bioassays with tobacco tissue cultures. Physiol. Plant. 15:473-497.

Randolph, L.F. 1936. Developmental morphology of the caryopsis in maize. J. Agric. Res. 53:881-915.

Rhodes, C.A., C.E. Green, and R.L. Phillips. 1982. Regenerable maize tissue cultures derived from immature tassels. Maize Genet. Coop. News Lett. 56:148-149.

Springer, W.D., C.E. Green, and K.A. Kohn. 1979. A histological examination of tissue culture initiation from immature embryos of maize. Protoplasma 101:269-281.

Methods of maize pollen germination in vitro, collection, storage, and treatment with toxic chemicals; recovery of resistant mutants

One vigorous corn plant sheds over 10^7 haploid, trinucleate pollen grains. Waxy and alcohol dehydrogenase-1 are known to be expressed after meiotic anaphase II, and the many correlations of duplication-deficient gametes with pollen abortion suggest that much of the pollen phenotype is encoded by its haploid genotype rather than the genotype of its pollen mother cell. For this reason, Nelson (commencing in 1958, Science 130: 794 with wx) and later, Freeling (1976, Genetics, in press) with Adh-deficient mutants among allyl alcohol-resistant pollen grains. Over the last two years, this laboratory has perfected numerous procedures involving maize pollen. Following Schwartz's lead, we have also recovered mutants via pollen selection. The methods and recipes we use follow. We hope they prove generally useful.

In vitro germination using "David's Bread Loaf": We have revised the pollen germination medium and conditions reported by Cook and Walden (1965, Can. J. Bot. 43: 779). Our concoction--called the Cook and Walden Revised Medium (CWRM)--is composed of 17% w/v sucrose, 300 mg/l $CaCl_2-2H_2O$, 100 mg/l H_3BO_3 and 0.7% w/v Difco Bactoagar, to a pH of 6.4 after the addition of agar but before heating. After heating until just clear (120°C for 8 min), this medium is gelled in a Griffin beaker of the desired diameter and stored at 4°C for at least four days without change. This column of gel is called "David's bread loaf, from which 2 mm slices are cut and immediately used for pollen germination at 25°C and low (uncontrolled) humidity within an unsealed petri dish. Our methods differ from Cook and Walden's in sucrose concentration, pH and humidity requirement; the major difference is that our pollen grains germinate on a newly cut solid surface. We achieve 75-95% germination of healthy pollen after 30 min for all of the seven different lines and inbreds we have tried. Before we devised the "bread loaf" technique, we experienced dramatic genotype-specific fluctuations; pH and sucrose concentration had to be continuously adjusted. The "bread loaf" technique affords the reproducibility necessary for determining kill curves for pollen pretreatments, and slices with gametophytes are easily moved to other dishes for staining, counting, fixing and the like.

Pollen collection and storage: The majority of our pollen viability studies utilized a 23± 2°C greenhouse, low humidity and natural January to March lighting. We find that our typical plant sheds over a five-day period. First day pollen usually germinates poorly; second and third day collections 2-4 hrs after dawn are optimal; we avoid afternoon collections. Tassels should be stripped of anthers and pollen the evening prior to collection. Pollen is collected as shed in glassine bags, desiccated ($CaSO_4$) and stored at 4°C for 30 min to 4 hr. This cold storage-desiccation consistently elevates percent gemination 5-10% to our modal 85%.

Berkeley's cool, dry summer permits routine collection of viable pollen from the field. Temperatures above 35°C or any discernible humidity greatly reduced the pollen's germinating ability. We typically collect 8×10^5 grains per plant in the late afternoon (2-5 p.m. PDT), subject them to chemical selective treatment, and pollinate at dusk. Our field pollen is from 50 to 90%

viable. We suspect that the heat and humidity characterizing a corn belt summer might necessitate using cooled greenhouses for the male parents.

Pollen counting: Pollen samples were suspended in 40% technical glycerin in a 250 ml graduate cylinder to a final concentration of 4-6 x 10^3 pollen grains per ml by visual estimation. A 1.0 ml sample of homogeneous suspension was further diluted and the sample was layered evenly over a gridded Gelman filter (GA-6, 0.45 um, 47 mm diam.) and quickly deposited by evacuation; circular currents were avoided in the layering process. The number of articles in four radial strips, each containing 5 squares (each square is one percent of the total area of the filter) were counted under incident light at a total magnification of 16X. In the rare instances where these numbers were significantly different from a 1:1:1:1: (by X^2), the entire sampling process was repeated. For each original pollen sample, two filters were prepared, counted and preserved for further reference; this gave eight statistically equivalent numbers on which to base our estimates of total pollen grains per ml of original suspension. In general, 1 mg of our desiccated pollen contains about 2,000 grains.

When simply counting for percent germination, the slice of germination medium is placed over a grid of any desired color, or over a transparent grid if underlighting is desired.

Treatment with toxic vapor and recovery of resistant mutants: We have selected pollen grains resistant to various toxic vapors. Only allyl alcohol (CH_2=CH-CH_2)H) resistance (rationale after Megnet, 1967, Arch. Biochem. Biophys. 121: 194, in yeast)--as it affects alcohol dehydrogenase activity--has been biochemically characterized: Megnet's scheme selectively kills ADH^+ cells owing to their capacity to oxidize relatively innocuous allyl alcohol to deadly acrolein. Megnet's selection is now being used in Drosophila and, as previously cited, Schwartz has reported success with maize ADH in pollen. Neither the selection scheme nor the strategy of using the male gametophyte are original to this laboratory, but the methods and results which follow are.

Pollen grains from a plant heterozygous for a gamma-induced ADH-deficient mutant (Adh-Fy25, Freeling, unpublished), Adh^+/Adh^- were collected, stored, and dispensed in 20 mg lots (40,000 grains) onto glassine paper for vapor treatment in a 500 ml Mason jar with sealed lid containing 20 ml of $CaSO_4$ dessicant. In this experiment, methanol was used as an inert carrier. 0 - 50 ul 100 allyl alcohol was pipetted into the jar before sealing (1 ul allyl alcohol vapor/500 ml is ca. 7.4 x 10^{-5} M). After 40 min of treatment, the pollen was evenly dusted with a camel hair brush over a newly cut slice of CWRM and germinated as described. After the germination was complete, the slice was frozen at -23^oC for 3 hr and defrosted for 30 min. ADH-specific stain was carefully layered over the gametophytes, left for 30 min, and replaced three times using methods of Freeling and Brown (1975, MGCNL 49:19). The 50% of the pollen grins that were ADH^+ stained blue and opague; the other 50%, ADH^-, stained yellow and translucent. Thus the four phenotypes: blue with pollen tube, blue without tube, yellow with tube, and yellow without tube. The data tabulated below compare the allyl alcohol with curves for sibling Adh^+ and Adh^- male gametophytes.

Allyl Alcohol		Mean Relative Percent Germination*	
ul/500 ml	mM	ADH+ (blue)	ADH- (yellow)
0	0	100 (78.5%)	100 (80.7%)
1	0.07	40	98
2	0.15	8.5	99
4	0.29	Zero	86
10	0.74	Zero	77.5
20	1.47	Zero	55.8
30	2.21	Zero	36.8
40	2.94	Zero	30.5
50	3.68	Zero	8.3

*The results of four independent experiments are averaged
for each data point; mean absolute control was 79%; allyl
alcohol was diluted with methanol such that 50 ul of
foreign vapor were present; treated for 40 min.

According to the data, treatment under these conditions with about 0.3 mM
allyl alcohol should let just a few Adh+ or Adh- sperm participated in the
fertilization. The results of these progeny tests were:

Allyl Alcohol (mM)	Seed Set	No. Seeds Receiving Allele*	
		Adh+	Adh-
0.30	100%	0	72
0.30	100	0	70
0.22	100	3	65
0 Control	100	42	38

*Less than 25% of the seed set was genotyped.

We conclude that ability to germinate in vitro at least approximates
ability to fertilize an ovum. However, with ADH+ pollen is treated with higher
concentrations of allyl alcohol, permitting fewer than 10^{-5} germinations, the
vat majority of these survivors remain ADH+. As might be expected, there are
alternative ways to be resistant to any toxin.

Pollen Propaganda: Since the recent dawn of "plant somatic cell genetics"
much has been said of selecting mutants in haploid, totipotent suspension
cultures. We submit that the tricks of "plant germ cell genetics" are not yet
exhausted. Pollen is available in huge numbers; is haploid; mutants may be
recovered in informationally normal cells; and pollen expresses many
differentiated functions, possibly permitting selection of agronomic traits.
We see the pollen grain a a powerful genetic resource. We hereby grant
permission to cite these methods.

David S.K. Cheng and Michael Freeling

NOTES ON CYTOGENETIC TECHNIQUES

I.   Cytological Smear Techniques
II. Methods for Pollen Sterility Determinations
III. Waxy Pollen Classification

C. R. Burnham
University of Minnesota
Revised in 1978

I. Cytological Smear Techniques

Carmine had been used for staining before 1850. Directions for making acetocarmine were given by Schneider in 1880 (Conn 1933).

The iron-acetocarmine smear technique was developed by Belling (1926) for use in studying the chromosomes at meiosis in plant microsporocytes. In one preparation, the chromosomes were well-stained; but the results could not be duplicated until he found that in the second trial he had used nickel-plated needles but had used rusty needles previously. Thus he discovered that iron was essential for proper staining of the chromosomes. Later, McClintock discovered that heating of the slides served to destain the cytoplasm and greatly improve the contrast between it and the chromosomes. Huskins stated that heating improved the accuracy of chromosome counts and analyses so much that he repeated much of his earlier work on speltoids in wheat. In addition to microsporocytes, the iron-aceto-carmine technique has been used successfully for chromosome morphology and counts on root tips, and embryonic leaf, bud and callus tissues in plants. Miss Stevens (1908) used aceto-carmine smears in studies of the chromosomes in _Drosophila_ testes. It has been used extensively for staining the chromosomes in salivary glands of _Drosophila_ and other _Diptera_; and on embryonic tissues of insects, birds and mammals; also for _Neurospora crassa_.

Many cytologists are now using _acetic-orcein_ and report better results than with aceto-carmine. Since there are several disadvantages in using it: it must be used on fresh material, it frequently precipitates; the beginner is advised to master the technique of making good slides using aceto-carmine before attempting the acetic-orcein method. For some materials orcein will stain the chromosomes, whereas aceto-carmine will not. According to Darlington and LaCour, acetic-orcein should not be used for tissues stored in 70% alcohol.

The brazilin stain (Belling 1930) made a rapid smear technique possible for _Oenothera_ chromosomes, for which aceto-carmine had not been satisfactory.

The aceto-carmine method will be described first based mainly on experience with _Zea mays_.

Aceto-carmine stain: preparation.

Method 1. Corn pachytenes require a very strong aceto-carmine. Carmine in
large excess of what will dissolve (about .5 gm per 100 cc) is added to boiling
45% acetic acid, and boiled for 1 to 2 min or until there is a sudden change to
a darker color. Cool, then filter. If not sufficiently dark, boil up again
with carmine.

Method 2. Using the same proportions as above, simmer the acetic acid and
carmine for 1--4 hr in a flask with a reflux condenser. Refluxing 6--8 hr
gives the best results for corn (Stout 1966). Cool and filter. This method
gives more consistent results and is recommended. Store the stock supply in a
brown bottle in the dark. It will keep indefinitely.

For use with barley and wheat dilute with an equal part of 45% acetic acid.

Propiono-carmine stain: preparation

Prepare as for acetocarmine, substituting 45% propionic acid for the 45% acetic
acid. Propionic acid dissolves more carmine stain than does acetic acid and
may give a clearer cytoplasm (McCallum, as given by Johannsen, 1940).

Fixing or killing solutions:

Farmer's fluid		Carnoy's fluids	
		A	B
3 parts	95% alcohol	6 parts	6
1 part	glacial acetic acid	3 parts	1
--	chloroform	1 part	3

These killers should be mixed fresh immediately before use. For corn Farmer's
fluid is satisfactory and 95% alcohol does as well as absolute alcohol. A
large excess of killer over the quantity of tissue may give better fixation,
especially when 95% alcohol is used. For some material absolute alcohol may be
better. Other proportions of alcohol to acetic acid may be better under
certain conditions or form different species. Carnoy's fluids are reported to
be better for wheat chromosomes. Some prefer the A mixture, others the B
mixture. Several in Great Britain add a few drops of a saturated aqueous
solution of ferric chloride to the killer after the material had been in the
killer for about 30 minutes or so. For the usual small buds or florets these
penetrate so rapidly that the outer floral parts need not be removed for better
fixing.

Satisfactory results were obtained in rye when methyl alcohol was substituted
for ethyl alcohol (Putt, 1954). This may be useful if ethyl alcohol is
difficult to obtain.

Propionic in place of acetic acid may give better fixation for some materials.

Temporary seals

A satisfactory temporary seal must prevent the slide from drying out, must not
smear the front lens of the microscope objective, and yet must be capable of
being removed easily in case the slide is to be made permanent.

1. Dahl's varnish-beeswax paraffin
    1/2 part Turtox Ringing Varnish
    1 part beeswax
    2 parts paraffin (parowax)

    This seems to make a very good airtight seal, satisfactory in all respects.

2. Gum-mastic, paraffin in equal parts are melted and stirred together. It
    may be necessary to strain the melted mixture through cheesecloth or wire
    screen to remove the hard material which is usually found in the gum
    mastic. Presence of this in the mixture might scratch the lens of the
    objective when changing from low to high power for figures near the edges
    of the cover slip. This mixture has a tendency not to stick to the slide
    if there is skin grease on it.

3. Lanolin-paraffin. Equal parts of each are melted together. This makes an
    airtight seal, but tends to be a little too soft, resulting in a tendency
    for the cover slip to slide when removing the seal in preparation for
    making it permanent, and also to leave a smear on the lens if it touches
    the seal. Possibly a mixture including beeswax would correct these faults.

4. Paraffin and vaseline--described by Baker (1952).

Details of the smear technique for studying chromosomes in plant species

These notes are based mainly on experience with Zea mays. For a review of the
method, see Smith 1947. Lima-de-Faria (1948) has described in detail his
methods for studying pachytene stages in rye.

1. Collection of sporocyte material: corn

    In corn the microsporocyte material for meiotic stages is taken usually
    about 5 to 10 days before the tassel tip first begins to make its
    appearance. The young developing tassel may be located inside the tightly
    rolled sheaths by squeezing the upper part of the plant between the thumb
    and forefinger. When this region feels soft and spongy the oldest portions
    are usually at or near meiosis. Another method of checking the stage of
    growth of the tassel is to cut or twist off the upper portions of the
    central whorl of leaves. When this exposes the tip of the tassel, the
    portions again are at or near meiosis. By checking cytologically,
    experience can be gained as to the appearance of the tassel, spikelets and
    anthers at the desired stages of meiosis.

    If well-developed tillers are available, or if pollen from the same plant
    is not required, the entire tassel may be taken. If the ear is needed,
    slit the stalk lengthwise with a razor blade to expose the tassel and cut
    the stalk off just below the base of the tassel. If the ear is not needed,
    pull up on the central portion of the whorl of leaves at the top of the
    plant. The stalk usually breaks at a point several internodes below the
    tassel. Remove the leaves and place the entire tassel in a vial of freshly
    mixed killer. Portions obviously too old can be discarded. A small cork
    attached to the end of a dissecting needle or pencil can be used
    to push the material into the killer, thus avoiding contact between one's

129

skin and killer. Repeated contact often causes the skin to peel.

If pollen from the same plant is desired, make a long vertical slit with a
razor blade through culm surrounding the spongy region to expose the young
tassel. Bending the plant slightly at that point opens the slit so that
the main branch or part of the side branches may be removed with tweezers
or fingers. Place the material directly into freshly mixed killer.
Removal of the branches without breaking facilitates locating the desired
meiotic stages later. The remaining portion of the tassel is pressed back,
the stalk straightened up to bring the edges of the slit together and the
stalk then is wrapped with masking tape to prevent drying out and to keep
the top of the plant upright as it grows. It may be necessary to tie the
top to a stake. Dusting with sulfur may be helpful if corn smut is a
problem. If properly done, the tassel will continue to develop so that it
can be used for pollinations.

Progression of stages in the tassel

The first anthers to shed pollen, and hence the oldest ones, are in
spikelets located a little above the middle of the central spike of the
tassel. The side branches below begin to flower a little later, and again
the first anthers to shed are a little above the middle. The spikelets are
progressively younger from that point toward the tip and toward the base of
each branch. If the central spike is collected when the oldest spikelets
are undergoing meiosis, the side branches usually will all be too young.
Collections made when anthers in the middle spikelets on the side branches
are at meiosis will furnish the greatest amount of material at that stage.
If the tassel is old, it is important to include the lower branches and all
florets to the base of each. Often in very old tassels, a few of the basal
florets will be young enough for stages of meiosis.

When beginning a new season of sporocyte collecting, check cytologically
for proper stage. The anthers often seem to be larger for a particular
stage than in later plants. It may be necessary to check for stage as the
season progresses.

Collection of sporocyte material from other species. In barley and in
wheat it is relatively easy to judge the stage by the length of head (about
1"-1 1/2" for diakinesis) in the boot. The head may be felt in the boot by
gentle pressure between the thumb and the forefinger. The oldest anthers
are in spikelets slightly above the middle of the head. If the plants are
well-tillered, it is usually possible to check the pollen for sterility and
then collect sporocytes from tillers of the desired plants.

For dicotyledons, it is necessary to judge the stage by the size of the
bud. For the first trial, collect a range of bud sizes, including
extremely small ones. Check the stages in each before collecting more
material.

Length of time in killer (fixative).

The anthers from freshly collected material may be smeared in acetocarmine
stain without previous killing. Only gentle heating can be used after
applying the cover slips. In corn, better results are obtained after the

130

sporocytes have been at least 24 hr in killer.

For pachytene study in corn, there is an optimum length of time to leave the material in the killer before preparing the slides in order to get the sharpest staining of the chromomeres; but this can be determined only by preparing slides at invervals after the material is collected. At room temperature the best slides with corn are usually made at from 12 to 24 hr after the material has been collected. Material which has been in killer at room temperature for more than 36 hr before making the smears usually does not give satisfactory pachytene preparations. High temperature or an excess of killer will decrease this time. When stored in the refrigerator immediately after collection, the material gives satisfactory pachytene slides for a much longer period of time.

Storage at temperatures below freezing (e.g. in a deep-freeze box) prolongs their usefulness still further (several years). To avoid poor fixation, place at that temperature only after they have been at room temperature for at least 24 to 36 hr after collection. Material that has been stored in the refrigerator should be allowed to stand at room temperature for a time (the optimum time to be determined as above by making slides at intervals) before making the smear preparations. For later use: for pachytene study, the killer should be replaced by 70% alcohol after 24 to 36 hr at room temperature, after 7 to 10 days in the refrigerator, after 1 to 2 months if below freezing. One change is sufficient if an excess of alcohol is added. Material in 70% alcohol possibly need not be stored in the refrigerator. Material in 70% alcohol stored in the refrigerator has given good pachytene preparations in maize after as long as 6 years, but the results are not as consistent, and some changes in technique are necessary to get the proper intensity of staining. For gross diakinesis, metaphase, or microspore studies the material may be left in the original killer. As a routine procedure, however, it is best to change all material to 70% alcohol even if stored below freezing.

To make up 70% alcohol from 95% alcohol use 70 cc of 95% and add 25 cc of distilled water. (Amount of water to add = 95-70 or 25 cc).

3. Preparation of the slide. Place the sporocyte material in a petri dish and add killer solution or 70% alcohol to prevent drying out. The black background of the laboratory table permits the anthers to be seen easier than if white filter paper is placed in the petri dish. Spread out the material so that one or two branches can be followed from end to end.

    a. Using rusty needles, when seeking diakinesis stages tease out the three long anthers of a single floret and place them in a drop of aceto-carmine. Anthers placed on a dry slide may shrivel before a drop of stain can be added and become very difficult to smear. As mentioned before, nickel-plated needles are not satisfactory. Cut the anthers transversely with a razor blade or between the tips of the needles. By gently tapping the anter pieces with the needle, the sporocytes are forced out of the open or cut ends of the anther without breaking them into small pieces. This makes it easier to remove the anther pieces before adding the cover slip. Stir the drop vigorously to separate the cells and also to incorporate iron into the stain. Examine the slide under low power to determine the stage (note that cells near the edge of the drop may be stained more densely, and the stage more easily recognized). Cell size and shape can be used to distinguish between separated microspores and pollen mother cells. Cells that have a very lightly stained central area are usually at pachytene or earlier. Ability to recognize the stage of meiosis without the cover slip will save time in slide preparation. See "General Notes" section for a detailed description of the progression of stages in the tassel.

b. When preparing slides for pachytene stages, tease out one anther and place it in a <u>small</u> drop of aceto-carmine stain. The drop should be small enough so that, when the cover slip is applied, the solution just barely fills out to the edges. It should <u>not</u> be necessary to apply pressure to squeeze out an excess of carmine on a blotter. If the drop is small, surface tension pulls the cover slip to the slide. An excess of stain also results in a slide in which the cells become too darkly stained after a few hours or overnight.

c. If at the desired stage, and stained densely enough (see "General Notes" section), remove the pieces of anther by <u>picking them up</u> between the points of the two needles. This seems to leave more cells in the drop than when the anther pieces are dragged off to one side. There are 3,000 to 4,000 pollen grains in an anther and hence 750 to 1,000 P.M.C.'s. A common problem is the loss of all but a few cells. Remove all dust particles in the area outside the drop, in the area to be covered, by brushing them off with the finger tip before applying the cover slip. Dust, lint or anther material in the drop will prevent the cover slip from being pulled down properly to the slide. If the drop is too large, remove a little of it with the finger tip. <u>Good slides cannot be made if the working space is dusty or if dirty slides or cover slips are used</u>. When the cover slip is added, if it fails to spread out at one or more spots, local light pressure with the point of the dissecting needle will aid in getting the carmine to spread and fill out to the edges of the slip. Care must be taken not to move the cover slip after the slip is applied. After applying the cover slip, do not try to remove the small trapped air pockets. They do not harm, and do serve as an indicator when the slide is being heated. Adding a minute amount of Dow's Anti-Foam Acid may aid in eliminating air bubbles under the cover slip (Yagyu, 1956, U. of Minnesota, Ph.D. thesis, on tomatoes).

In preparing slides for <u>diakinesis</u> stages, a larger drop of stain may be used. After removing the anther pieces and adding the cover slip, excess stain may be removed by inverting the slide on a blotter and applying gentle pressure, being certain not to move the cover slip.

d. <u>After adding the cover slip, heat over an alcohol flame</u>. Pass the slide (middle portion) back and forth over an alcohol flame, allowing it to heat to a point just short of boiling. Examine under low power of the microscope to note the change in differentiation of the stain; the cytoplasm becomes lighter and the chromosomes more densely stained. Repeat the heating until there is no further change in differentiation. Slides may need to be heated as many as 10-15 times. If they boil, the cells and chromosomes are ruined, but some very good figures have been found on slides that had boiled at the edges. Some

132

have used a heating plate, or a steam bath, but it is important to watch the process under the microscope.

A test often seen is that of touching the back of one hand with the heated slide. This is not satisfactory, since the proper temperature is one that will burn the hand. In corn, with fresh material or material that has not been in killer long enough, heating has a tendency to remove the protoplasm, leaving only the chromosomes. Only gentle heating can be used.

e. Apply the temporary seal, covering the top edges of the cover slip as well as the sides to make certain it is completely sealed.

f. With a wax pencil, record the culture and plant number on the slide, as well as the stage.

If made properly, the slides will keep for some time, but will gradually become dark or the stain may crystallize out. An excess of stain in the original preparation results in a more rapid deterioration. The slides may keep longer in a refrigerator, although they are more likely to dry out unless kept in the hydrator pan or petri dish with moisture.

A variation of the above technique which results in slides which will keep longer before being made permanent is used by Dr. A.E. Longley: Just before adding the cover slip, i.e. after the chromosomes are well stained, a small drop of glacial acetic acid is added at the edge of the drop of carmine and stirred into it gradually. Then the cover slip is added, and heat applied. The larger drop size presents some difficulty, and may have to be squeezed out on a blotter. This method appears to result in a better destaining of the cytoplasm and in slower crystalizing of the carmine as the slide ages. He used a larger, rectangular cover slip.

General Notes

Tassel morphology in relation to locating desired stages.

In corn, the spikelets are often paired, the sessile one often being at a slightly earlier stage than the adjacent long-stalked one. Within each spikelet, there are two flowers, each with three anthers, those in the first or upper flower being much longer and hence older than those in the lower or second flower. Anthers from the first flowers in the spikelets show the greatest regularity in succession of stages along the branch. Also the three anthers in the first flowers are usually at about the same stage of meiosis. This is very useful in studying pachytene stages. If half an anther is used to make a slide, as many as 6 slides at a given stage may be made. This increases the chance of getting some well-made preparations from that flower.

Under some conditions, not all the stages are on a single branch, the jumps being too great between successive spikelets. The missing stages may be found on a different branch when this occurs, the three anthers of one flower may not be at the same stage. If all the first flowers are too old, the desired stages may be found in the second flowers with the shorter anthers. In these, the succession is usually not as regular, i.e. not in the expected order from one spikelet to the next. In some cases where diakinesis is desired, a few cells

133

in this stage may be present in an anther in which most of the cells are at metaphase I. (For a description of the staminate inflorescence, see Hayward, 1938, pp. 130-132).

Miscellaneous Notes

In maize the spore-mother cell wall in most cases has slipped off the cell either in the killing or in the smearing process. The empty shells are frequently seen in the smear. The nuclear membrane is removed also by the killer. Both enhance the spreading of the cell contents.

The sporocytes are easily broken if the anther is crushed or if pressure is applied to the cover slip. For this reason spear-point needles are not recommended. It is best to prepare your own needles, replacing the commerical dissecting ones by a size 7 needle or smaller. The pointed end of the needle should be curved by bending it in a hot flame. The curved ends make them easier to manipulate removing excess anther material.

In some cases slides made immediately after removing the material from the refrigerator will be free from "bubbles" in the cytoplasm, whereas slides made later may have them.

Since the configurations in permanents may not be as good as the freshly stained ones, all observations and photographs are made before making the slides permanent.

Intensity and color of the stain is influenced by the following and may therefore be controlled by varying these factors:

1. Intensity of the aceto-carmine solution.

2. The amount of iron added by the iron needles. This is determined by the amount they are used in stirring and teasing the material (also the degree of rustiness of the needles).

   Another method is to add to the killer a few drops of a solution of ferric hydrate dissolved in 45% acetic acid. Barton (1950) for tomatoes used a series with increasing amounts of iron. As mentioned earlier, some add ferric chloride to the killer. For pachytene study, too much iron stains the nucleolus so heavily that chromosomes above or below it cannot be studied.

3. Oxidation: Longer exposure to the air will intensify the staining. If it is still too light in color after applying the cover slip, the intensity may be increased by allowing the slide to stand for a time before heating the slide.

Method for making centromeres clearly distinguishable
(McClintock, 1957, personal communication)

"Preparations are made in the usual way with carmine. The slides should stand for at leat four days or a week. Then, the seal on two opposite sides should be removed and a stain made from acetic acid, lactic acid and orcein should be drawn under the cover slip, replacing the carmine stain. (The stain: 1 part water, 1 part acetic acid, one part lactic acid and 2% orcein). After a day, the orcein stain will have replaced the carmine stain. The DNA-containing parts of the cell are then deeply stained and the other parts are very lightly stained. In maize, this makes the positions of the centromeres very conspicuous as they are not stained." She reports this gives excellent results in maize.

An improved aceto-carmine smear technique for sporocytes that spread poorly
(Neubauer, 1966b)

Corn anthers for pachytene, diakinesis or metaphase I analysis are removed from the acetic-alcohol killer (or from 70% alcohol if already changed) and placed for a few minutes in 20% acetic acid before proceeding with the regular staining procedure. This usually results in better spreading of the chromosomes. The diameter of the cells was increased 70 to 80%. Similar, but less pronounced effects were observed at metaphase I in barley. Prolonged exposure results in loss of affinity for stain.

Cleaning cover slips

Cover slips can be cleaned in $K_2Cr_2O_7$ + sulfuric acid cleaning solution, but the immersion oil must be removed first (apply a few drops of xylene, wipe off with cheesecloth). The cleaning solution plus immersion oil forms a gummy mass.

Making aceto-carmine or propiono-carmine smears permanent (McClintock, 1929)

Remove the temporary seal with a razor blade, off the top of the cover slip first, then along the sides; removing the remainder with a xylol-moistened camel's hair brush. Examine the slides; if the cytoplasm is too dark, run 45% acetic acid under the cover slip by adding a drop of acid along one side, pulling it through with a blotter held against the opposite side. Then heat to destain, repeat until destained satisfactorily.

Place right side up in a petri dish of 10% acetic acid. The solution should soak under the cover slip and loosen it. If not, then while pressing the spread ends of the forceps gently against one side or edge of the cover slip to keep it from sliding, run the edge of the razor blade under the opposite side to raise the cover slip from the slide. Keeping the cover slip and slide in the same relative position they originally had, run both through a series of solutions listed below. The original position of the cover slip is usually indicated by a line of aceto-carmine crystals on the slide, or its positions may be marked on the bottom side with a diamond pencil. If carried out properly the figures will be found very near their original vernier readings.

135

Pass cover slip and slide through the following freshly prepared solutions in coplin jars, about 2 minutes each:

1. 10 cc 95% alcohol + 30 cc of 45% acetic acid
2. 20 cc "   "   + 20 " "   "   "   "
*3. 20 cc absolute alcohol + 20 cc of glacial acetic
4. 30 cc   "   "   + 10 " "   "   "
*5. 36 cc   "   "   + 4 " "   "   "
*6. 40 cc absolute alcohol
*7. 40 cc   "   "

*These steps alone as a short method have been satisfactorily form corn and for barley.

Then add balsam of proper consistency to the slide and quickly place the cover slip in its original position, using a needle to shift it about. Turn the slide bottom side up. Check cover slip position again. Then place on a piece of paper toweling or a blotter and press out the excess balsam, being careful not to move the cover slip out of position. Transfer the slide to a fresh piece of paper towel, still bottom side up. Flood with xylol, being certain the towel under the slide is saturated. When dry, the slide will be relatively free of excess balsam. My experience with clarite has been that one application is not sufficient to seal to cover slips, unsealed areas appear as the slide dries.

General notes on the method: Slides that are a few days old seem to lose fewer cells than ones made into permanents immediately. Most of the sporocytes adhere to the cover slip, very few are on the slide.

When step #7 with absolue alcohol was omitted, the edges of the cells frequently were darker or folded over; possibly due to moisture since this additional step (#7) with absolute alcohol has corrected this.

In a moist atmosphere a milky-white percipitate soon forms on the surface of the drop of balsam; hence the need for adding the cover slip quickly.

The alcohol-acetic acid mixtures in the series probably remain usable longer if stored in a refrigerator.

A step using pure xylol preceding balsam might logically be expected, but McClintock found that it results in inferior slides. It was only when the xylol became mixed with alcohol from the preceding steps that the slides were satisfactory. No xylol is used in the method now.

Other methods of making permanents have been proposed. Sears' (1941) short tertiary butyl alochol method is satisfactory but not better for corn and barley than the method presented above. Longley's method is to soak off the cover slip in a mixture of equal parts of 95% ethyl alcohol and glacial acetic acid, followed by two changes of absolute alcohol and then mounting in Euparal.

Bridges' (1937) alcohol vapor method, which consists of placing the unsealed slide in a sealed jar with several strips of absorbent paper saturated with absolute alcohol and leaving them for 24 to 48 hr; has resulted in a general darkening of the cytoplasm in corn. Its advantage for Drosophila salivary

136

chromosomes is that the cover slip is not removed from the slide.

Quick-freeze method of making slides permanent, as described by Conger and Fairchild (1953).

Place the temporary slide on the flat side of a block of dry ice. When thoroughly frozen and while still on the ice, pry off the cover slip by inserting a razor blade under one corner of the cover slip. The cells adhere to the slide. Place the frozen slide immediately in 95% or absolute alcohol. Remove the slide, without draining, add a very large drop of Euparal at an edge of the cell-bearing area. With the edge of the cover slip, draw the Euparal up over the cells and gently lower the cover slip. Do not press out the excess mounting medium for a day or so. The slow penetration of the Euparal into the alcohol layer over the cells apparently prevents the collapse of the cells. The advantages are: ease and speed of removing the cover slip with a minimum loss of cells, superiority of the permanent slides, and vernier readings of cell positions for the temporary mount remain the same for permanent preparation.

We have used balsam as the mounting medium, to avoid the extra manipulation with Euparal. The two should be compared. Baker (1952) reports its successful use for Drosophila salivary glands and ganglion chromosome preparations.

For the study of the nucleoli in microspore quartets, 1. more iron must be worked into the drop than is used for sporocytes. After the nucleoli are densely stained, the cytoplasm may be destained by applying heat or a combination of the Longley method (addition of a small drop of glacial acetic acid) plus heat.

Cooper's staining method: the use of aceto-carmine to which 1/4 part of Ehrlich's haematoxylin has been added will also give darker staining, especially of the nucleolus (weak carmine solutions may be made fairly usable for sporocytes in this manner). Delay in heating after applying the cover slip will also result in darker-staining nucleoli. If the first division plane is to be determined, a larger-sized drop should be used, and heat applied cautiously.

2. Propionic acid-cotton blue for holding spore quartets together.
J. Neubauer (1966a)

        25 ml distilled, deionized water
       125 ml glycerin
        25 gm (carbolic acid) phenol
        25 ml lactic acid
        .1 gm cotton blue

After mixing, boil very slowly until 1/4 of the solution is boiled away. After cooling, mix 1 part of stain with 2 parts of propionic acid.

If spore quartets well-stained with propiono-carmine are mounted in the above mixture, the spores remain in the original spore mother cell wall. Some destaining is possible if steam heat is used. If the quartets reject destaining, use less propionic cotton blue. If destaining is too drastic, use more of the mixture.

137

For first division of microspores in corn

An anther is tested into aceto-carmine in a hollow-ground hanging-drop slide or in a watch glass, and allowed to stain for 20-30 min, adding more stain as it evaporates. A small drop of the suspension may be transferred with a wire loop to a slide, cover slip added, heat applied, and sealed. In corn, the pollen grains show a characteristic furrowed or wrinkled appearance when at the proper stage. This is usually at about 7 to 10 days after meiosis, or when the tip of the tassel first begins to show. With corn, the percentage of microspores in division stages is much higher in greenhouse than in field grown plants.

For wheat microspores, Bhowal (1963) describes the following technique:

1. Anthers at the suitable stage in a vial of saturated aqueous solution of paradichlorobenzene (A Merk) are held at 10-15 C for 6 hr to overnight.

2. Fix in Carnoy B fixative over night at room temperature.

3. Hydrolyze in N HCl at 60°C for 13 min.

6. Rinse only once in distilled water (Pectinase works best in an acid medium).

7. Treat with 1% aqueous solution of pectinase (Nutritional Biochemical Corp.) at room temperature for 1/2 hr.

8. Rinse well with distilled or tap water.

9. Stain in basic fuchsin for 1-2 hr.

10. Squash anthers in 1% aceto-carmine, remove debris, add cover slip, place slide between blotting paper, apply gentle rolling pressure, slightly warm the slide and seal.

The enzyme powder should be stored in a refrigerator and should not be older than about 1 year.

Smear techniques for other species

Barley and wheat require a more dilute aceto-carmine and less iron. One that is best for corn will result in black cells in barley. In both barley and wheat, after the cover slip is applied, place the slide between layers of blotter on the table and apply pressure with the thumb or a print roller.

Drosophila. The salivary glands are allowed to fix for 20-30 minutes in the stain before smearing. Glass needles are usually used to give slower crystallization of the aceto-carmine.

Neurospora crassa. Perithecia were placed in Carnoy's B fixative over night, and stored at 1°C. Cells were stained with aceto-carmine, the slides inverted on blotting paper and considerable pressure applied to flatten the asci (Phillips, 1967).

<u>Techniques</u> <u>used</u> <u>by</u> <u>Stringam</u> <u>for</u> <u>tomato</u>

Two methods of fixation gave superior staining of spore mother cells:

<u>Method 1</u>.   (Swaminathan et al. 1954)
                 2 parts absolute alcohol
                 1 part of a saturated solution of ferric acetate in propionic
acid

<u>Method 2</u>.   (particularly useful for buds collected from older
                 plants)
                 2 parts absolute alochol
                 1 part of a mixture of:
                     5 cc of propionic acid
                     1 drop of 5% ferric salt of EDTA

Buds with sufficient absorption of iron were dark green in color. "Sporocytes from buds that remained white after fixation by the first method, usually were readily stained following their transfer to 70% alcohol to which had been added about 1 drop of the EDTA solution per cc and storage for about 24 hours in the refrigerator." After 1 to 2 weeks, transfer to 70% alcohol and store at about 0 F. Good staining was obtained even after a year.

Anthers were squashed in a small drop of propiono-carmine, anther pieces removed, and cover slip added. Heating of the slide over steam from a small flask for 15 to 20 seconds was essential for good staining and differentiation of the chromosomes (Stringam, 1966).

<u>Melilotus</u>. Methods similar to the above (Kita et al. 1959) for tomatoes gave satisfactory results. Method 1 of fixation was used.

<u>Onenohera</u> <u>and</u> <u>other</u> <u>species</u> <u>with</u> <u>woody</u> <u>anthers</u>
(J. Stout, 1966)

Oenothera anthers have a woody texture, but the following procedure macerates the tissue. Ordinary procedures used for aceto-carmine smears do not stain the chromosomes satisfactorily. This corrected by the method of fixation used for tomatoes:

1. Fix in 2 parts absolute ethanol: 1 part propionic acid saturated with ferric acetate (same as for tomato) for minimum of 4 hr.

2. Transfer to a 1% aqueous solution of "Clarese 300" (Miles Chemical Company division of Miles Laboratories, Inc., Elkhart, Indiana) for 4 or more hours at room temperature.

3. Remove material and store in 70% ethanol.

4. Stain in the usual way with aceto- or propiono-carmine allowing iron from needles to darken the stain.

Various enzymes have been used--see Emsweller and Stuart, 1944; Faberge, 1945; Bachman and Bonner, 1959; Sharman, 1960.

Smears of root tips or of other growing tissues

Root tip smear method as used by Hagberg for barley
(Tjio and Levan 1950, modified by Hagberg)

<u>Solutions:</u>

1. 8 hydroxyquinoline (145 = molecular weight) - that from Merck Chemical Co., Darmstadt, Germany, is satisfactory. That available in the U.S. does not give good results. (The two differ in appearance also).

    Make up a solution in distilled water, 0.004 molar (nearly saturated) - i.e. 0.145 gm in 250 cc. For most materials this can be diluted 1/2 to .002 molar (standard), some species need it stronger.

    Warm to 60 C for 10-15 min or until all dissolved; as much as 1 hr is O.K., but not overnight.

2. <u>Acetic orcein stain preparation</u>

    Orcein source: G.T. Gurr, London. Again, the U.S. source behaves differently (more sensitive to acid), and also requires a greater dye concentration.

    Dissolve 4.4 gm orcein in 100 cc glacial acetic acid, reflux about 30 min by gentle boiling, cool, filter. Dilute when needed for use (it deteriorates in dilute acid). Dilute about 1/2 with distilled water to a strength of 45% acetic acid---this gives about a 2% orcein solution. This is used to make up the macerating mixture.

    It is further dilute 1/2 (with 45% acetic acid) to make up the 1% acetic orcein which is used for staining.

    <u>Technique</u>, as applied to barley

    1. Start the seedling in moist sand in petri dishes at room temperatures. Sand gives better results than filter paper---breakage studies gave different results on sand vs. filter paper.

    2. When about 1 cm long* (short tips are best), remove and place in oxyquinoline solution which is kept at about 15_0C (cooler than summer temperatures)---running tap water in our lab is satisfactory, 4 hr.

    3. Then transfer to a watch glass with a mixture of:
        9 part of the 2% orcein in 45% acetic acid
        1 part Normal HCl (10 cc of conc. HCl per 100 cc is close to N HCl).

    4. Heat gently for 5-10 sec, 3 to 4 times, not too hot to touch. Leave at least five min. If difficult to macerate, add 1 or 2

drops of N HCl to the watch glass <u>after</u> the first heating. At the first heating the HCl must not exceed 1:9 or stickiness of the chromosomes results.

5. Transfer the root tip to filter paper to remove the acid mixture.

6. Transfer to slide, and with a spear point needle, cut off the <u>very</u> short meristematic end, add an excess of 1% acetic orcein.

7. Add the cover slip. Then, holding one edge of the cover slip down so it cannot move, tap the cover slip with a rubber-tipped pencil 2 or more times. Then with a folded filter paper covering about 2/3 of the cover slip apply <u>heavy</u> pressure with the thumb, being very careful not to move the cover slip, rolling the thumb across the cover slip, applying the pressure as it is rolled. The slide is ready for examination. If properly done, the cells in division should have separated from the remainder of the root tip tissue (this is controlled by the tapping with the lead pencil); and the plates that are in metaphase should show the chromosomes somewhat shortened and well flattened. For further flattening, local pressure may be applied with a spring clothespin on a paper wad placed directly over the figure (Caldecott). It may be best to get a photograph before this is done, as well as after.

Plumule growing plants have more divisions and may be better if only counts are needed.

*If early enough, can get 1st division of the embryonic root tip.

<u>Modifications used by Ramage for barley root tips</u>

For step #2, use a solution of 0.1% colchicine. Leave standing in light for 20-30 min, place in refrigerator for 24 hr.

3. Remove and place in Carnoy's fluid A for 24-72 hr.

4. Remove and place in 1 part 95% alcohol:1 part conc. HCl for 5-10 min.

5. Remove and place in fresh Carnoy's A for 5-10 min.

6. Place root tip on slide in drop of aceto-carmine, cut off meristematic end of root and discard remainder.

7. Chop the tip into small bits with a razor blade. Use of a small stirring rod with a smooth rounded or flattened end was very effective. If material is not to be smeared immediately it should be transferred to 70% alcohol after 24 to 36 hr in the original killer.

8. Add the cover slip and gently heat. <u>While warm</u>, place slide between several sheets of filter paper and apply pressure to spread and flatten the cells.

9. Heat until desired differentiation is obtained.

For material that does not stain well with aceto-carmine, Brown has used for maceration (in step #2 above) the following mixture:

1 part 95% alcohol
1 part concentrated HCl
2 parts of a 4% aqueous solution of iron ammonium sulfate (iron alum)

The other steps in the process are the same.

Methods used by Tuleen for barley root tips

Seeds were planted in 3-inch pots. About 1 week after planting, rootips were collected and placed in a vial containing a .002 M solution of 8-hydroxyquinolin (.073 gm in 250 cc). It is best to add the tips to a refrigerated solution. Some have frozen vials in an ice cube tray for field collections. The root tips were refrigerated in this solution at 36-40°F for 4 to 8 hr to shorten the chromosomes, then transferred to a 1 N solution of HCl and hydrolyzed at 140°F for about 10 min. They were then transferred to a vial containing basic feulgen stain and stored at 36-40°F. The root tips were squashed in a drop of acetocarmine. Material stored in feulgen longer than 2 or 3 days tended to deteriorate (Tuleen 1966).

Preparation of Leuco-basic fuchsin modified formula after de Tomasi and Coleman as given by Darlington and La Cour (1947)

1. Dissolve 1 gm basic fuchsin by pouring over it 200 cc boiling distilled water.

2. Shake well and cool to 50°C.

3. Filter, add 30 cc N HCl to filtrate.

4. Add 3 gm $K_2S_2O_5$.

5. Allow solution to bleach for 24 hr in a tightly stoppered bottle in the dark; add .5 gm decolorizing carbon (Norit, a vegetable carbon, is recommended).

Technique for red clover root tips (using HCl-carmine stain). (Snow 1963, with modification by J. Neubauer, and K.C. Nag)

Preparation of HCl-carmine Stain

Add 3 gm carmine and 2 ml conc. HCL to 15 ml H_2O in glass beaker. Boil slowly for 10 min and stir constantly. Allow it to cool. Add 85 ml of 80% ethanol and stir. Filter and collect the stain.

1. Fix root tips in Farmer's solution (3 parts 95% Ethanol and 1 part glacial acetic acid) for 24 hr at about 35°F, do not freeze.

2. Rinse root tips with 95% alcohol for 10 min.

3. Stain in HCl-carmine stain for 48 hr at room temperature.

4. Rinse in 45% acetic acid for about 2 min.

5. Rinse in 70% ethanol for 15 min.

6. Rinse in 95% ethanol for 5 min.

7. Store in 75% ethanol in refrigerator, do not freeze.

Slide preparation:

Take stained root tip and clean it (i.e. make it free from dirt, etc.) and put it on a drop of Propiono-carmine stain on the slide. Macerate the root tip by means of sharp needles. Then press the macerated material by the smooth flat "butt-end" of a pair of forceps. This pressing helps in the spreading of the cytoplasmic material. A proper preparation will show the chromosomes very clearly and well spread in the cell. Add cover slip at this time.

Pre-treatment with paradichlorobenzene

Paradichlorobenzene has an effect similar to that of colchicine, i.e. it shortens the chromosomes and prevents spindle formation. This results in more countable figures, whereas without treatment most of the cells show side views of metaphase, usually not countable. The method as proposed by Meyer (1945) with some modifications by O'Mara (personal communication) is the following:

The prefixative is a saturated solution prepared by mixing 5-10 gm paracichlorobenzene in 500 cc of distilled water in a stoppered bottle and holding it at about 60°C over night. Use at room temperature.

1. Remove rapdily growing root tips and place in the prefixative. The time may vary from 15 min to 4 hr depending on the species--in corn, Retherford used 2 to 3 hr.

2. Pour off prefixative, add fixative and proceed as with other methods.

Grass leaf chromosomes

A saturated solution of bromonaphthalene in .05% saponin solution was used for pre-treatment. The chemical was more soluble in the saponin solution than in water. The best contraction of the chromosomes was after 50 to 90 minutes (Bennett, 1964).

Hill and Myers (1945) found pre-treatment with cold resulted in shorter somatic chromosomes in certain forage grasses.

Rapidly growing leaves and stem tips

For phlox, Meyer (1943) 1. removed young leaves and stem tips with a razor blade and immersed them in a 0.2% colchicine solution. 2. They were placed under a light (60-100 watts) for 1 to 2 hr. Do not let the material wilt before immersing. Leaves 1 mm long may have actively dividing cells in certain species, in others 4 mm long. Remove older, outer leaves to expose the youngest leaves. Blot the colchicine solution from the material. Fix material (young leaves and stem) in Semmen's Carnoy (3 parts absolute alcohol:1 part glacial acetic acid: 1 part chloroform), or acetic-alcohol. Proceed with the usual staining metods.

II. Determination of pollen sterility

The pollen of corn or barley heterozygous for various chromosomal aberrations is partially sterile. The visible abnormal grains may be completely empty (devoid of starch) and smaller than the normal ones, partially filled with starch, or well-filled but smaller in size; the proportions varying with the aberration.

For checking in the field, a "Leitz Taschen Mikroskop" with a magnification of 40X is the most convenient. It is about 1" in diameter, 1 1/2" long and uses a small glass slide which is inserted in a slot below the lens which can be focussed by screwing it in or out. It is no longer manufactured by Leitz, but an instrument similar mechanically to this, and with only slightly less magnification, called the Midgard at $5 each is manufactured by:

> Nippon Micropscope Works Company
> 35-2 Minami Cho
> Aoyama, Akasaka
> Tokyo, Japan

The opening in the bottom of the instrument can be enlarged with a metal reamer to let in more light (Burnham, 1961).

The magnification is not sufficient for classification of species with smaller pollen, such as barley and tomato.

Field classification: An anther freshly extruded or about to extrude is removed, the tip pinched off with a fingernail and the anther rolled between the fingers to dust the pollen on the small glass slide. Young anthers with mature pollen, but not ready to shed can be classified by crushing and smearing the anther on the slide with a finger; but this is too slow for general use. Fertile plants were marked by tearing off all but about 2" of the top 2 or 3 leaves, partially sterile plants were marked with a strung tag on which the degree of sterility could be recorded. Where there was only one partially sterile class, with rare exceptions, a 2' long string was looped around the plant.

For accurate counts in the laboratory a 5- or 6-inch section of tassel which includes a portion that has not sheed but also a portion that has shed (thus assuring the inclusion of spikelets with mature anthers) is tagged with the culture and characters, the different phenotypes can be collected and tied in separate bundles, one branch being taken from each plant.

To determine the degree of visibly aborted pollen, the pollen from an anther about ready to shed is teased out into a drop of I_2-KI solution and examined under the microscope. Experience gained by estimating the degree of abortion and then checking the same slide by counting enables one to make fairly accurate estimates at least sufficient to distinguish the major classes of 25,

50, or 75%, and with practice the classes with values midway between them.

By screwing off the front lens of the low power objective sufficient magnification is obtained (32 or 40X) and with a larger field which aids in making estimates. A special 4X, 32 mm low power objective will give a flatter field.

For making accurate counts, 1/3 of an anther is teased out into a very small drop of I_2-KI solution, and a small cover slip (narrow strip, about 1/3 of the 7/8" square) added. This may be ringed with mineral oil to prevent drying out during the counting.

A circular disc of note card cut to fit on the shelf in the ocular and with a rectangular opening cut out as in the diagram: gives a field with parallel sides. Strips across the slide are counted by recording the total of a single class in each trip across the strip. Since the aborted grains tend to float to the outside the entire slide is counted. With a 6-bank counter, the classes in each field can be counted.

Strong I_2-KI solution
0.3 gm I_2
1.0 gm KI
100 cc H_2O

If glycerin is added to the solution, the grains will stay in place better (Pittenger and Frolik)

This may be diluted several times for the pollen sterility classification.

Tomato. The platform plus the upper portion of an old monocular microscope, with the base removed, was fitted with a carrying strap for use in the field. Pollen was dusted on a slide for examination.

Other Methods

1. Aceto-carmine is used by those working with potatoes to determine the percentage of "stainable" or normal pollen. In most cases, the two methods probably give the same results; although in certain cases they may not (Leasley and Lesley, J. Agr. Res. 58:621-630. 1939).

2. Aniline blue in lactophenol (E.D. Garber). Add enough dye to give a dense solution. The slides keep.

Material should be checked with the different methods.

III. Classification for waxy

For classification of pollen for waxy vs. non-waxy, the stock solution should be diluted with three parts of water. Tease out an anther in a drop on a slide, examine with the light cut off from the mirror but with light shining on

145

it from above (a method discovered by McClintock).

For classifying waxy in sugary seeds, use the undiluted solution or one with 0.6 gm of I_2 per 100 cc of water.

In barley seeds, the classification is more difficult than in corn. In barley, the red-brown color shows only for an instant.

General references

1. Consult Stain Technology regularly for new techniques (now in Agric. Library).

2. Belling, J. 1930. The Use of the Microscope pp. 243-5 (McGraw-Hill). Brazilin stain for material difficult to stain with aceto-carmine.

3. Bronte, Gatenby, and Beams. 1950. 11th edition. Lee's Microtomist's Vade-Mecum.

4. Conn, H.J. 7th ed. 1961. Biological Stains. Williams and Wilkins, Baltimore.

5. Conn, H.J. 1933. The History of Staining. Geneva, N.Y. Book Service of the Biological Stain Commission.

6. Darlington, C.D. and L.F. LaCour. 1947. The Handling of Chromosomes. Allen & Univin, London.

7. Johansen, D.A. 1940. Plant Microtechnique. McGraw-Hill, N.Y.

Other references, many for techniques not included in this writeup.

Aass, Inger. 1957. (A cytological analysis of Scots pine, Pinus sylvestris L.) Meddel. Norsk. Skogfors. (Reports of Norvegian Forest Res. Inst. 48 (14):96-109. Eng. Abs. Needle primordia used.

Bachman, B. and D.M. Bonner. 1959. Protoplasts from Neurospora crassa. Jour. Bacteriology 78:550-556. (Use of snail enzyme preparation to liberate intact living protoplasts).

Baker, Wm. K. 1952. Permanent slides of salivary and ganglion chromosomes. Dros. Inf. Service (C.S.H.) No. 26:129.

Baldwin, J.T., Jr. 1938. Kalanchoe: the genus and its chromosomes. Amer. Jour. Bot. 25:572-579. (A modification of Warmke's method).

Barton, D.W. 1950. Pachytene morphology of the tomato chromosome complement. Amer. Jour. Bot. 37:639-643.

Belling, J. 1926. The iron-aceto-carmine method of fixing and staining chromosomes. Biol. Bull. 50:160-162.

Bennett, Erna. 1964. A rapid modification of de Latour's technique for grass leaf chromosomes. Euphytica 13:44-48.

Bhowal, J.G. 1963. Use of pectinase in the study of pollen mitosis of wheat. Canad. J. Gen. and Cytol. 5:268-269.

Bridges, C.B. et al. 1937. The vapor method of changing reagents and of dehydration. Stain Tech. 12:51.

Brown, Wm. L. 1937. A modified root tip smear technique. Stain Techn. 12:137-138.

Brown, S.W. 1949. The structure and meiotic behavior of the differentiated chromosomes of the tomato. Genetics 34:437-461. (Describes a technique developed by Marta S. Walters for the tomato).

Buck, J.B. 1935. Permanent aceto-carmine preparations. Science 81:75.

Burnham, C.R. 1961. Notes on the pocket microscope. Maize Gen. Coop. News Letter 35:88.

Burrell, P.C. 1939. Root tip smear method for difficult material. Stain Techn. 14:147-9.

Conger, A.D. and L.M. Fairchild. 1953. A quick-freeze method for making smear slides permanent. Stain Tech. 28:281-283.

Datta, P.C. and A. Naug. 1967. Staining pollen tubes in the style; cotton blue vs. aceto carmine for general use. Stain Tech. 42:81-85 (tested on 4 monocots, 32 dicots).

Emsweller, S.L. and N.W. Stuart. 1944. Improving smear technique by the use of enzymes. Stain Techn. 19:109-14.

Faberge, A.D. 1945. Snail stomach cytase, a new reagent for plant cytology. Stain Techn. 20:1-4.

Flagg, R.O. 1961. Prolonged storage of Feulgen preparations in water. Stain Tech. 36:95-97. Better chromosome scattering in taxa with high numbers and higher % of good figures in root tip squashes.

Hayward, H.E. 1938. The Structure of Economic Plant. MacMillan, N.Y.

Heddle, John A. 1967. Graphical conversions of mechanical stage readings for field findings in different microscopes. Stain Tech. 42:109-111.

Heilborn, O. 1937. A new method of making permanent smears with special reference to salivary gland chromosomes of Drosophila. Annals Agric. College of Sweden 4:89-98. (Landbrukshogskolens Annaler).

Heilborn, O. 1940. Further contributions to a chromosome analysis of Lilium. Hereditas 26:100-106.

Hill, H.D. and W.M. Myers. 1945. A schedule including cold treatment to facilitate somatic chromosome counts in certain forage grasses. Stain Tech. 20:89-92.

Hillary, B.B. 1938. Permanent preparations from rapid cytological technics. Stain Tech. 13:161-7.

Hillary, B.B. 1939. Improvements in the permanent root tip squash technique. Stain Techn. 14:97-9.

Hillary, B.B. 1939. Use of the feulgen reaction in cytology. I. Effect of fixatives on the reaction. Bot. Gaz. 101:276-

1940. II. New techniques and special applications. Bot Gaz. 102:225-35.

Jona, R. 1963. Squashing under Scotch tape No. 665 for auto-radiographic and permanent histologic preparations. Stain Tech. 38:91-95.

Jona, R. 1967. Handling germinated pollen on millipore membrane during the Feulgen and autoradiographic procedures. Stain Tech. 4:113-117.

Kaufman, B.P. 1927. The value of the smear method for plant cytology. Stain Tech. 2:88-90.

Kita, F., M.L. Magoon, and D.C. Cooper. 1959. Simple smear techniques for the study of chromosomes of *Melilotus*. Phyton (Argentina) 12:35-38.

LaCour, L. 1941. Acetic-orcein: a new stain-fixative for chromosomes. Stain Techn. 16:169-174.

Lima-de-Faria, A. 1948. B chromosomes of rye at pachytene. Portugaliae Acta Bibl. Gen. A 2:167-173.

Marengo, U.P. 1967. The relationship of microtome knife facet bevels to edge effectiveness. Stain Tech. 42:23- Describes adjustments for AO automatic knife sharpener which allow 4 sections to be cut.

Matsuura, H. 1938. A simple new method for the demonstration of spiral structure in chromosomes. Cytologia 9:243-8.

McClintock, Barbara. 1929. A method for making acetocarmine smears permanent. Stain Techn. 4:53-6.

Meyer, J.R. 1943. Colchicine-Feulgen leaf smears. Stain Techn. 18:53-56.

Meyer, J.R. 1945. Prefixing with paradichlorobenzene to facilitate chromosome study. Stain Tech. 20:121-5.

Neubauer, J. 1966a. Propionic acid cotton blue (for spore quartets). Maize Gen. Coop. News Letter 40:106.

Neubauer, J. 1966b. An improvement in aceto-carmine smear technique. Ibid. 40:106-107.

Ostergren, G. and A. Bajer. 1958. Permanent preparation from endosperm cells flattened in the living state. Hereditas 44:466-470.

Painter, T.S. 1939. An aceto-carmine method for bird and mammalian chromosomes. Science 90:307-8.

Phillips, R.L. 1967. The association of linkage group V with chromosome 2 in Neurospora crassa. Jour. Hered. 58:263-265.

Pittenger, T.H. and E.F. Frolik. 1950. Preparing slides for determining percentage of pollen abortion. Maize Genetics Coop. News Letter #24:60-61.

Putt, E.D. 1954. Cytogenetic studies of sterility in rye. Canad. J. Agric. Sci. 34:81-119.

Savage, J.R.K. 1967. Double staining for comparative measurements in squash preparations. Stain Tech. 42:19-21.

Sax, Karl. 1931. The smear technic in plant cytology. Stain Tech. 6:117-122.

Sears, E.R. 1941. Chromosome pairing and ferility in hybrids and amphidiploid in the Triticinae. Mo. Agr. Exp. Sta. Res. Bul. 337:1-20.

Sharman, B.C. 1960. Volvox colonies and snail cytase. Nature 186:90. (failed to separate the cells).

Smith, Luther. 1947. The acetocarmine smear technique. Stain Tech. 22:17-31.

Smith, S.G. 1943. Techniques for the study of insect chromosomes. Canadian Entomologist 75:21-34.

Snow, R. 1963. Alcoholic hydrochloric acid-carmine as a stain for chromosomes in squash preparations. Stain Tech. 38:9-13.

Steere, W.C. 1931. A new and rapid method for making permanent aceto-carmine smears. Stain Tech. 6:107-111.

Stevens, N.M. 1908. A study of the germ cells of certain Diptera. Journ. Exp. Zool. 5:359-374. (testes or ovaries of adult flies dissected out in physiological saline and transferred to a drop of acetocarmine on a slide.
Press the cover glass down with a needle to break the capsule of the testis and spread the cells. (Remove excess stain with filter paper).

Storey, W.B. and J.D. Mann. 1967. Chromosome contraction by o-isopropy N-phenylcarbamate (IPC). Stain Techn. 42:15-18 (this is a selective herbicide).

Stout, J. 1966. Improved propiono carmine stain. Maize Gen. Coop. News Letter 40:107.

Stringam, Gary R. Cytogenetic studies of interchanges involving chromosome 2 in the tomato, <u>Lycopersicon</u> <u>esculentum</u> Mill. Ph.D. Thesis, Univ. of Minn. 1966.

Swaminathan, M.S., M.L. Magoon, and K.L. Mehra. 1954. A simple propionic-carmine P.M.C. Smear method for plants with small chromosomes. Indian J. Gen. Plant Breeding 14:87-88.

Swanson, C.P. 1940. The use of acenaphthene in the pollen tube technic. Stain Techn. 15:49-52.

Taylor, W.R. 1924. The smear method for plant cytology. Bot Gaz. 78:236-238.

Thomas, P.J. 1940. The acetocarmine method for fruit material. Stain Techn. 15:167-72.

Tjio, J.A. and A. Levan. 1950. The use of oxyquinoline in chromosome analysis. An. Aula Dei. 2:21:64.

Tuleen, Neal A. The use of tertiary trisomics for the orientation of linkage groups in barley. Ph.D. Thesis, Univ. of Minn. 1966.

Warmke, H.E. 1935. A permanent root tip smear method. Stain Techn. 10:101-103.

Zirkle, C. 1940. Combined fixing, staining and mounting media. Stain Techn. 15:139-53.